BIOTECHNOLOGY
INTELLIGENCE
UNIT

Tissue Repair, Contraction and the Myofibroblast

Christine Chaponnier, Ph.D.
Department of Pathology and Immunology
University of Geneva
Geneva, Switzerland

Alexis Desmoulière, Pharm.D., Ph.D.
Inserm E362
University of Bordeaux 2
Bordeaux, France

Giulio Gabbiani, M.D., Ph.D.
Department of Pathology and Immunology
University of Geneva
Geneva, Switzerland

LANDES BIOSCIENCE / EUREKAH.COM
GEORGETOWN, TEXAS
U.S.A.

SPRINGER SCIENCE+BUSINESS MEDIA
NEW YORK, NEW YORK
U.S.A.

TISSUE REPAIR, CONTRACTION AND THE MYOFIBROBLAST

Biotechnology Intelligence Unit

Landes Bioscience / Eurekah.com
Springer Science+Business Media, LLC

ISBN: 0-387-33649-4   Printed on acid-free paper.

Copyright ©2006 Landes Bioscience and Springer Science+Business Media, LLC

Springer Science+Business Media, LLC, 233 Spring Street, New York, New York 10013, U.S.A.
http://www.springer.com

Please address all inquiries to the Publishers:
Landes Bioscience / Eurekah.com, 810 South Church Street, Georgetown, Texas 78626, U.S.A.
Phone: 512/ 863 7762; FAX: 512/ 863 0081
http://www.eurekah.com
http://www.landesbioscience.com

Printed in the United States of America.

9 8 7 6 5 4 3 2 1

Library of Congress Cataloging-in-Publication Data

Chaponnier, Christine.
    Tissue repair, contraction, and the myofibroblast / Christine Chaponnier, Alexis Desmoulière, Giulio Gabbiani.
        p. ; cm. -- (Biotechnology intelligence unit)
    Includes bibliographical references and index.
    ISBN 0-387-33649-4 (alk. paper)
    1. Wound healing. 2. Myofibroblasts. 3. Regeneration (Biology) I. Desmoulière, A. (Alexis), 1957- II. Gabbiani, Giulio. III. Title. IV. Series.
    [DNLM: 1. Wound Healing--physiology. 2. Extracellular Matrix--metabolism. 3. Fibroblasts--pathology. 4. Fibrosis--physiopathology. QZ 150 C462t 2006]
    RD94.C43 2006
    617.1'4--dc22
                            2006008951

CONTENTS

Preface xi

Acknowledgements xii

Introduction—The Evolution of the Concept of Myofibroblast:
Implications for Normal and Pathological Tissue Remodeling 1
Alexis Desmoulière, Christine Chaponnier and Giulio Gabbiani
    Fibroblast/Myofibroblast Transition 2
    Role of α-SM Actin in Tension Generation 2

1. Cytomechanics in Connective Tissue Repair and Engineering 7
Robert A. Brown
    Stress-Shielding of Resident Cells against Applied Forces 8
    Force Vectors and ECM Anisotropy 10
    The Central Problem of Scarring Is Collagen
        Matrix Contracture 10
    Growth Repair and Contracture of the ECM:
        What Do They Involve? 11
    What Then Are the Real Functions of Fibroblast
        Force Generation? 16
    Cell-Molecular Responses to Cytomechanical Cues 17
    Contraction and Contracture for Collagen
        Remodelling (Shortening) 20

2. Scleroderma Lung Fibroblasts:
Contractility and Connective Tissue Growth Factor 25
*Galina S. Bogatkevich, Anna Ludwicka-Bradley, Paul J. Nietert
and Richard M. Silver*
    Scleroderma Lung Fibrosis and Myofibroblasts 25
    TGF-β, Thrombin and CTGF in SSc Lung Fibrosis 26
    Contractile Activity of CTGF 27

3. Functional Assessment of Fibroblast Heterogeneity
by the Cell-Surface Glycoprotein Thy-1 32
*Carolyn J. Baglole, Terry J. Smith, David Foster, Patricia J. Sime,
Steve Feldon and Richard P. Phipps*
    Immunological and Inflammatory Characteristics of Thy-1
        Fibroblast Subsets 33
    Fibrogenic and Proliferative Characteristics of Thy-1[+]
        and Thy-1[-] Subsets 35

4. Tissue Repair in Asthma: The Origin of Airway
    Subepithelial Fibroblasts and Myofibroblasts 40
    Sabrina Mattoli
    Phenotypic and Functional Characteristics
      of Circulating Fibrocytes 41
    Phenotypic Characteristics and Bone Marrow Origin
      of Tissue Fibrocytes 41
    Identification of Circulating Fibrocytes as Precursors
      of Bronchial Myofibroblasts in Asthma 42
    Mechanisms Potentially Involved in the Recruitment
      of Fibrocytes into the Airways in Asthma 43

5. Experimental Models to Study the Origin
    and Role of Myofibroblasts in Renal Fibrosis 47
    Michael Zeisberg, Mary A. Soubasakos and Raghu Kalluri
    COL4A3-Deficient Mice 47
    Nephrotoxic Serum Nephritis 48
    Unilateral Ureteral Obstruction 49
    Which Is the Best Model to Use? 49
    Histopathology and Morphometric Analysis 50

6. Epithelial to Mesenchymal Transition of Mesothelial Cells
    as a Mechanism Responsible for Peritoneal Membrane
    Failure in Peritoneal Dialysis Patients 53
    Abelardo Aguilera, Luiz S. Aroeira, Marta Ramírez-Huesca,
    José A. Jiménez-Heffernan, Rafael Selgas and Manuel López-Cabrera
    Peritoneal Fibrosis 54
    Role of TGF-β in the Pathogenesis of Peritoneal Fibrosis 56
    Implication of Epithelial-Mesenchymal Transition
      of MC in Peritoneal Fibrosis 58
    Role of Epithelial-Mesenchymal Transition of MC
      in Neovascularization and Peritoneal Transport Disorders 59
    Therapeutic Intervention on EMT 62

7. FIZZy Alveolar Epithelial Cells Induce
    Myofibroblast Differentiation 68
    Sem H. Phan
    FIZZ1 Expression 69
    Effects of FIZZ1 on Fibroblasts and Myofibroblast Differentiation 70
    Regulation of FIZZ1 Expression 71
    FIZZ1 in Pulmonary Fibrosis 71

8. Pro-Invasive Molecular Cross-Signaling between Cancer
   Cells and Myofibroblasts 74
   Olivier De Wever and Marc Mareel
   Host Cells Participate at Cancer Cell Invasion 74
   Myofibroblasts Stimulate Invasion 75
   Cancer Cell-Derived TGF-β Converts Fibroblasts
      into Myofibroblasts 77
   Myofibroblasts Are Pro-Invasive through the Combined
      Action of Tenascin C (Tn-C) and Hepatocyte Growth
      Factor/Scatter Factor (HGF/SF) 79
   Myofibroblasts Are Themselves Invasive 82
   The Pro-Invasive Switch in the Cross-Signaling Pathway 82

9. Proangiogenic Implications of Hepatic Stellate Cell
   Transdifferentiation into Myofibroblasts Induced
   by Tumor Microenvironment 88
   Elvira Olaso, Beatriz Arteta, Clarisa Salado, Eider Eguilegor,
   Natalia Gallot, Aritz Lopategi, Virginia Gutierrez, Miren Solaun,
   Lorea Mendoza and Fernando Vidal-Vanaclocha
   Cancer Microenvironment and Tumor-Activated
      Myofibroblasts 89
   Pathophysiologic Aspects of the Hepatic Metastasis Process 90
   Hepatic Stellate Cell Transdifferentiation
      into Myofibroblasts during the Microvascular
      Stage of the Hepatic Metastasis Process 90
   Hypoxia Induces Proangiogenic Activation of Hepatic
      Stellate Cell-Derived Myofibroblasts in Avascular
      Micrometastases: Implications on Intratumoral
      Endothelial Cell Recruitment and Survival 94
   Structural Relationships between Myofibroblastic
      and Neo-Angiogenic Patterns of Developing
      Hepatic Metastasis 95
   Intrametastatic Myofibroblasts Support Metastasis
      Development via Paracrine Cancer Cell Invasion
      and Proliferation-Stimulating Factors 96
   Targeting Tumor-Associated Myofibroblasts as a Novel
      Approach to Anti-Tumor Treatment in the Liver 97

10. **Matrix Metalloproteinases, Tissue Inhibitors of Metalloproteinase and Matrix Turnover and the Fate of Hepatic Stellate Cells** 102
Aqeel M. Jamil and John P. Iredale
   A Brief Review of the Role of Activated Stellate
      Cells/Myofibroblasts in Hepatic Fibrosis 103
   The Regulation of Hepatic Stellate Cell Apoptosis 104
   Soluble Cytokines and Survival Factors in the Regulation
      of Stellate Cell Apoptosis 105
   The Role of TNF Receptor Super Family Members
      in Mediating Stellate Cell Apoptosis and Survival 105
   Matrix Stability and the Role of Tissue Inhibitor
      of Metalloproteinases in Mediating Stellate Cell Survival 105

11. **Innate Immune Regulation of Lung Injury and Repair** 110
Dianhua Jiang, Jennifer Hodge, Jiurong Liang and Paul W. Noble
   Host Responses in Lung Injury 110
   Hyaluronan and CD44 in Lung Injury and Repair 111
   Interferon-γ (IFN-γ) and CXCR3 in Lung Fibrosis 113

12. **An Eye on Repair: Myofibroblasts in Corneal Wounds** 118
James V. Jester
   Corneal Imaging Using in Vivo CM 119
   Wound Contraction following Incision Corneal Injury 120
   Cellular Mechanism of Wound Contraction in the Cornea 123
   Myofibroblasts, Tissue Growth and Corneal Haze 129
   TGF-β and Appearance of Myofibroblasts in Corneal Wounds 132
   Differentiation of Keratocytes to Myofibroblasts 133

Index 139

EDITORS

Christine Chaponnier
Department of Pathology and Immunology
University of Geneva
Geneva, Switzerland
Introduction

Alexis Desmoulière
Inserm E362
University of Bordeaux 2
Bordeaux, France
Introduction

Giulio Gabbiani
Department of Pathology and Immunology
University of Geneva
Geneva, Switzerland
Introduction

CONTRIBUTORS

Abelardo Aguilera
Unidad de Biología Molecular
Instituto Reina Sofía
   de Investigaciones Nefrológicas
Hospital Universitario de la Princesa
Madrid, Spain
Chapter 6

Luiz S. Aroeira
Unidad de Biología Molecular
Instituto Reina Sofía
   de Investigaciones Nefrológicas
Hospital Universitario de la Princesa
Madrid, Spain
Chapter 6

Beatriz Arteta
Department of Cell Biology
   and Histology
Basque Country University
   School of Medicine and Dentistry
Leioa, Bizkaia, Spain
Chapter 9

Carolyn J. Baglole
Department of Environmental Medicine
University of Rochester
   School of Medicine and Dentistry
Rochester, New York, U.S.A.
Chapter 3

Galina S. Bogatkevich
Medical University of South Carolina
Charleston, South Carolina, U.S.A.
Chapter 2

Robert A. Brown
Tissue Repair and Engineering Centre
Royal National Orthopaedic Hospital
   Campus, Stanmore
University College London
   Institute of Orthopaedics
London, U.K.
Chapter 1

Olivier De Wever
Department of Radiotherapy
University Hospital Ghent
Ghent, Belgium
Chapter 8

Eider Eguilegor
Dominion Pharmakine Ltd.
Bizkaia Technology
Derio, Bizkaia, Spain
Chapter 9

Steve Feldon
Department of Ophthalmology
University of Rochester
    School of Medicine and Dentistry
Rochester, New York, U.S.A.
Chapter 3

David Foster
Department of Obstetrics
    and Gynecology
University of Rochester
    School of Medicine and Dentistry
Rochester, New York, U.S.A.
Chapter 3

Natalia Gallot
Dominion Pharmakine Ltd.
Bizkaia Technology
Derio, Bizkaia, Spain
Chapter 9

Virginia Gutierrez
Dominion Pharmakine Ltd.
Bizkaia Technology
Derio, Bizkaia, Spain
Chapter 9

Jennifer Hodge
Department of Medicine
Section of Pulmonary
    and Critical Care Medicine
Yale University School of Medicine
New Haven, Connecticut, U.S.A.
Chapter 11

John P. Iredale
Liver Research Group
Southampton University
    School of Medicine
Southhampton General Hospital
Southhampton, Hants, U.K.
Chapter 10

Aqeel M. Jamil
Liver Research Group
Southampton University
    School of Medicine
Southhampton General Hospital
Southhampton, Hants, U.K.
Chapter 10

James V. Jester
Department of Ophthalmology
University of California at Irvine
Irvine, California, U.S.A.
Chapter 12

Dianhua Jiang
Department of Medicine
Section of Pulmonary
    and Critical Care Medicine
Yale University School of Medicine
New Haven, Connecticut U.S.A.
Chapter 11

José A. Jiménez-Heffernan
Departamento de Patología
Instituto Reina Sofía
    de Investigaciones Nefrológicas
Hospital Universitario de Guadalajara
Guadalajara, Spain
Chapter 6

Raghu Kalluri
Department of Medicine
Center for Matrix Biology
Harvard Medical School
Boston, Massachusetts, U.S.A.
Chapter 5

Jiurong Liang
Department of Medicine
Section of Pulmonary
   and Critical Care Medicine
Yale University School of Medicine
New Haven, Connecticut U.S.A.
Chapter 11

Aritz Lopategi
Department of Cell Biology
   and Histology
Basque Country University
   School of Medicine and Dentistry
Leioa, Bizkaia, Spain
Chapter 9

Manuel López-Cabrera
Unidad de Biología Molecular
Instituto Reina Sofía
   de Investigaciones Nefrológicas
Hospital Universitario de la Princesa
Madrid, Spain
Chapter 6

Anna Ludwicka-Bradley
Medical University of South Carolina
Charleston, South Carolina, U.S.A.
Chapter 2

Marc Mareel
Department of Radiotherapy
   and Nuclear Medicine
University Hospital Ghent
Ghent, Belgium
Chapter 8

Sabrina Mattoli
Avail Biomedical Research Institute
and
Avail GmbH
Basel, Switzerland
Chapter 4

Lorea Mendoza
Dominion Pharmakine Ltd.
Bizkaia Technology
Derio, Bizkaia, Spain
Chapter 9

Paul J. Nietert
Department of Biostatistics,
   Bioinformatics and Epidemiology
University of South Carolina
Charleston, South Carolina, U.S.A.
Chapter 2

Paul W. Noble
Department of Medicine
Section of Pulmonary
   and Critical Care Medicine
Yale University School of Medicine
New Haven, Connecticut, U.S.A.
Chapter 11

Elvira Olaso
Department of Cell Biology
   and Histology
Basque Country University
   School of Medicine and Dentistry
Leioa, Bizkaia Spain
Chapter 9

Sem H. Phan
Department of Pathology
University of Michigan Medical School
Ann Arbor, Michigan, U.S.A.
Chapter 7

Richard P. Phipps
Department of Environmental Medicine
University of Rochester School
   of Medicine and Dentistry
Rochester, New York, U.S.A.
Chapter 3

Marta Ramírez-Huesca
Unidad de Biología Molecular
Instituto Reina Sofía
   de Investigaciones Nefrológicas
Hospital Universitario de la Princesa
Madrid, Spain
Chapter 6

Clarisa Salado
Dominion Pharmakine Ltd.
Bizkaia Technology
Derio, Bizkaia, Spain
Chapter 9

Rafael Selgas
Servicio de Nefrología
Instituto Reina Sofía
   de Investigaciones Nefrológicas
Hospital Universitario La Paz
Madrid, Spain
Chapter 6

Richard M. Silver
Department of Medicine
Medical University of South Carolina
Charleston, South Carolina, U.S.A.
Chapter 2

Patricia J. Sime
Division of Pulmonary
   and Critical Care Medicine
University of Rochester
   School of Medicine and Dentistry
Rochester, New York, U.S.A.
Chapter 3

Terry J. Smith
Division of Molecular Medicine
Harbor-UCLA Medical Center
Torrance, California, U.S.A.
Chapter 3

Miren Solaun
Dominion Pharmakine Ltd.
Bizkaia Technology
Derio, Bizkaia Spain
Chapter 9

Mary A. Soubasakos
Department of Medicine
Harvard Medical School
Boston, Massachusetts, U.S.A.
Chapter 5

Fernando Vidal-Vanaclocha
Department of Cell Biology
Basque Country University
   School of Medicine and Dentistry
Bizkaia, Spain
Chapter 9

Michael Zeisberg
Department of Medicine
Harvard Medical School
Boston, Massachusetts, U.S.A.
Chapter 5

PREFACE

Thirty-four years after the first description of the myofibroblast, the number of publications concerning this cell is very impressive and continuously expanding, and the work on the myofibroblast involves many laboratories throughout the world. The myofibroblast has been implicated in developmental and physiological phenomena, as well as in a variety of pathological situations, going from wound healing and fibrotic changes to asthma and cancer invasion. Many aspects of myofibroblast biology have been clarified, such as the role of TGF-β and ED-A cellular fibronectin in its differentiation and the role of α-smooth muscle actin in tension production by this cell; however several important problems concerning myofibroblast origin, function and participation in pathological processes remain to be solved.

The purpose of this book, as well of the Meeting "Tissue Repair, Contraction and the Myofibroblast" that took place in Nyon, near Geneva, Switzerland on November 18-20, 2004, is to put together the most recent advances in the understanding of myofibroblast biology and to present the main directions of research taking place worldwide to explore new aspects of myofibroblast physiological and pathological activities, such as: mechanisms of force generation by the myofibroblast; myofibroblast origin and diversity; interaction of the myofibroblast with other cells, normal and malignant epithelial cells in particular; and participation of the myofibroblast in the development of fibrosis in various organs. If we consider the animated and constructive discussions that took place during the Nyon Meeting, we are sure that this book will inspire new research in these fields.

This book would not have existed without the help of the European Tissue Repair Society and the Swiss National Science Foundation as well as the several Sponsors who are listed in the acknowledgments.

We hope that it will be the first of a long and fruitful series.

Christine Chaponnier, Ph.D.
Alexis Desmoulière, Pharm.D., Ph.D.
Giulio Gabbiani, M.D., Ph.D.

Acknowledgments

We are very grateful to the Swiss National Science Foundation (Grant No 31CO00-107296/1 and Grant No IB71A0-108533) and to the European Tissue Repair Society for their financial support. We gratefully acknowledge the following companies and institutions for their support: Main Sponsors: L'Oréal Recherche (Paris, France), and Pierre Fabre Dermo-Cosmétique (Toulouse, France). Other Sponsors: Roche Pharma (Neuilly-sur-Seine, France), Carl Zeiss (Feldbach, Switzerland), PromoCell (Heidelberg, Germany), Urgo (Chenôve, France), Ville de Nyon (Switzerland), Région Aquitaine (France).

A special thanks to Jean-Claude Rumbeli for photographic and graphic computer work, to Evelyne Homberg for secretarial help, to the Department of Pathology and Immunology, and to the Faculty of Medicine, CMU-University of Geneva, Switzerland, for support.

INTRODUCTION

The Evolution of the Concept of Myofibroblast:
Implications for Normal and Pathological Tissue Remodeling

Alexis Desmoulière, Christine Chaponnier and Giulio Gabbiani*

Abstract

The recognition of the role of the myofibroblast in granulation tissue contraction and connective tissue remodeling during fibrocontractive diseases has allowed a theoretical and practical progress in the understanding of these pathologies. The observation that TGF-β is the key cytokine in myofibroblast differentiation, correlated with its role in collagen synthesis promotion, shows a coordinated mechanism in connective tissue remodeling. Recent work has furnished new knowledge concerning the molecular mechanisms of tension production by the myofibroblast and indicated that the N-terminal peptide of α-smooth muscle actin exerts an inhibitory action on myofibroblast contraction. Moreover the multiple derivation, both local and from circulating cells, of the myofibroblast begins to be understood. These data point to the myofibroblast as a major regulator of connective tissue remodeling and in turn of epithelial organization.

Introduction

After the first description of the myofibroblast in granulation tissue of an open wound by means of electron microscopy, as an intermediate cell between the fibroblast and the smooth muscle cell (SMC),[1] the myofibroblast has been identified both in normal tissues, particularly in locations where there is a necessity of mechanical force development (for a review, see ref. 2), and in pathological tissues, in relation with hypertrophic scarring, fibromatoses and fibrocontractive diseases (for a review, see ref. 3) as well as in the stroma reaction to epithelial tumors (for a review, see ref. 4). More recently myofibroblasts have been described in the deep dermis of patients with systemic sclerosis (for a review, see ref. 5) and in the bronchial submucosa of asthmatic patients (for a review, see ref. 6).

In an attempt to verify whether the myofibroblast expresses markers of the SMC phenotype, our laboratory has shown that α-SM actin, the actin isoform typical of vascular SMCs, is synthesized during fibroblast/myofibroblast modulation.[7] Indeed the presence of this protein represents at present the best marker of the myofibroblastic phenotype (for a review, see ref. 8).

*Corresponding Author: Giulio Gabbiani—Department of Pathology and Immunology, CMU-University of Geneva 1 rue Michel-Servet, 1211 Geneva 4, Switzerland. Email: Giulio.Gabbiani@medecine.unige.ch

Tissue Repair, Contraction and the Myofibroblast, edited by Christine Chaponnier, Alexis Desmoulière and Giulio Gabbiani.

When present in the myofibroblast, α-SM actin localizes in stress fibers, an organelle that has been at first considered characteristic of cultured cells, but is also present in vivo, particularly during myofibroblast differentiation. Myofibroblasts can also express other proteins characteristic of SMCs, such as SM-myosin heavy chains, according to the pathological situation;[9] however they have not been shown up to now to express late markers of SMC differentiation, such as smoothelin.[10] This allows distinguishing between the two phenotypes.

Fibroblast/Myofibroblast Transition

The mechanisms of fibroblast/myofibroblast transition have been the object of intensive investigation. Transforming growth factor (TGF)-β is now accepted as the most important factor in this transition since it stimulates both the synthesis of collagen type I[11] and of α-SM actin by fibroblastic cells.[12,13] Connective tissue growth factor has also been proposed to play a role in myofibroblast differentiation.[14] It is now accepted that during the healing of an open wound, fibroblast/myofibroblast transition begins with the appearance of the protomyofibroblast, which contains stress fibers expressing only cytoplasmic β- and γ-actin isoforms.[10] This first transition is not yet well explored, but it probably depends on mechanical tension development. Then follows the appearance of the differentiated myofibroblast under the influence of mechanical tension as well as of chemical mediators, such as TGF-β (Fig. 1). It should be noted that the action of TGF-β in stimulating both collagen type I and α-SM actin synthesis strictly depends on the presence of cellular fibronectin and in particular of the ED-A splice variant of this glycoprotein.[15] Thus myofibroblast differentiation is a complex process, regulated by at least a cytokine, an extracellular matrix component as well as the presence of mechanical tension (Fig. 1).

During the healing of an open wound, when epithelial reconstruction is achieved, an important wave of apoptosis is observed in the underlying granulation tissue affecting small vessel cells (endothelial cells, pericytes) and myofibroblasts, thus leading to the formation of scar tissue.[16] The lack of apoptosis has been suggested as one of the mechanisms involved in the development of hypertrophic scars and possibly of other fibrotic changes. However this possibility has not yet been thoroughly explored.

The local derivation of fibroblastic cells from preexisting fibroblasts during wound healing has remained a dogma since the early work of Ross et al.[17] Subsequent work by several laboratories has shown that indeed local fibroblasts are a major source of myofibroblasts; however myofibroblasts can derive also from local mesenchymal cells such as SMCs, pericytes, hepatic stellate cells or mesangial cells.[18] The derivation of myofibroblasts from SMCs is particularly interesting in view of the recently described different mechanisms of contraction of these two cells (see below). In the last years the possibility that myofibroblasts derive from local epithelial cells or from blood bone marrow derived cells, which was suggested very early in the literature (for review see refs. 17,18), has been again convincingly proposed. Thus, it appears that tubular epithelial cells of the kidney are at least in part the source of myofibroblasts during interstitial fibrosis[19] and that mesothelial cells can originate myofibroblasts during peritoneal fibrosis.[20] Moreover it is more and more accepted that a variable proportion of myofibroblasts present in different pathological situations, e.g., liver[21] and pulmonary[22] fibrosis, are bone marrow derived. In this respect it is noteworthy that the description of circulating cells, called fibrocytes, which localize in areas of repair[23-25] and are probably an important source of myofibroblasts.[26] Clearly the identification and characterization of such cells may have important implications for the understanding of reparative and fibrotic changes of many organs and for the planning of therapeutic strategies.

Role of α-SM Actin in Tension Generation

As discussed above, α-SM actin is the most used marker of the myofibroblastic phenotype (for a review, see ref. 27). However the question as to whether this protein is instrumental in force production by the myofibroblast has been debated for a long time. Recently our

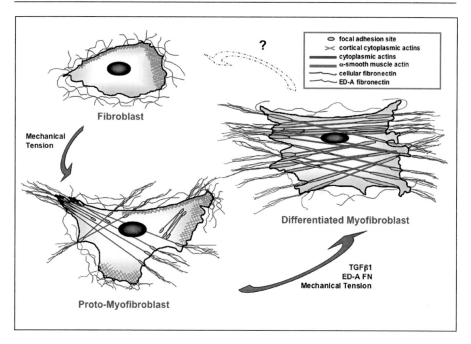

Figure 1. Two-stage model of myofibroblast differentiation. In vivo, fibroblasts might contain actin in their cortex but they neither show stress fibres nor do they form adhesion complexes with the extracellular matrix. Under mechanical stress, fibroblasts will differentiate into proto-myofibroblasts, which form cytoplasmic actin-containing stress fibers that terminate in fibronexus adhesion complexes. Proto-myofibroblasts also express and organize cellular fibronectin—including the ED-A splice variant—at the cell surface. Functionally, these cells can generate contractile force. TGF-β1 increases the expression of ED-A fibronectin. Both factors, in the presence of mechanical stress, promote the modulation of proto-myofibroblasts into differentiated myofibroblasts that are characterized by the de novo expression of α-smooth muscle actin in more extensively developed stress fibres and by large fibronexus adhesion complexes (in vivo) or supermature focal adhesions (in vitro). Functionally, differentiated myofibroblasts generate greater contractile force than proto-myofibroblasts, which is reflected by a higher organization of extracellular fibronectin into fibrils. (From ref. 10, ©2002 Nature Publishing Group, with permission.)

laboratory has shown that there is a good correlation between α-SM actin expression and the capacity of producing deformations in the silicone substrate on which fibroblastic cells are cultured; moreover transfection of swiss 3T3 fibroblasts with α-SM actin cDNA results in an increased contractility, which is significantly higher compared to that of fibroblasts transfected with the cDNA of α-cardiac actin or γ-cytoplasmic actin.[28] The increase in contractility takes place in the absence of any other change in protein expression, in particular of myosin heavy chain expression. These results strongly suggest that α-SM actin plays a direct role in tension production by myofibroblasts.

Two other observations have helped in pinpointing the mechanism of α-SM actin participation in tension production by myofibroblasts: (1) the decrease of the critical concentration for α-SM actin polymerization by the Fab fragment of the specific antibody for this protein, suggesting that binding of α-SM actin epitope facilitates incorporation of the protein into filaments of stress fibers;[29] and (2) the identification of the epitopic sequence for this antibody, i.e., the N-terminal sequence AcEEED.[29] The identification of a putative compound that in the cell would bind α-SM actin in a way similar to that of the antibody and thus increase its incorporation into stress fibers has not yet been possible, but we have shown that microinjection in cultured myofibroblasts of the epitopic sequence decreases significantly and selectively α-SM actin incorporation into stress fibers. The sequence Ac-EEED is very acidic and does not penetrate spontaneously in cells. In order to perform more systematic studies, we have coupled it with an Antennapedia sequence that facilitates cell penetration.[30] We have seen that such fusion peptide inhibits myofibroblast contractility both in vitro and in vivo, using an experimental model of splinted wound in the rat that facilitates the study of wound contraction by eliminating the role of epithelialisation.[31] These observations open the possibility to influence wound contraction and/or myofibroblast dependent connective tissue remodeling in vivo and thus may be the basis for a new therapeutic strategy concerning several connective tissue diseases in which the myofibroblast appears a key player.

The fusion peptide should also represent a useful tool for the understanding of the molecular mechanisms regulating myofibroblast driven connective tissue remodeling. It is now accepted that wound contraction, as well as probably contracture formation, depends on the continuous long lasting production of isometric tension by single myofibroblasts (for a review, see ref. 32). The resulting connective tissue retraction can be stabilized by extracellular matrix deposition. This complex dynamic process could explain connective tissue remodeling in normal wound healing and in pathological situations such as liver cirrhosis or pulmonary fibrosis. Recent work aimed to understand the mechanism of force generation by stress fibers has indicated that tension production by the myofibroblast is regulated differently with respect to the classical Ca^{++} depending reversible SM contraction and is rather under the control of a Rho/Rho kinase and myosin phosphatase related pathway.[33,34] These findings establish for the first time a clear difference between myofibroblast and SMC in terms of contraction mechanisms and suggest that the myofibroblast utilizes a more primitive mechanism of force production that bears some analogies with the extracellular matrix remodeling taking place during embryonic development.[35] In this respect it is noteworthy that myofiboblasts have been described in embryonic tissues of various species, including man,[36] but their possible participation to developmental phenomena has not been explored. If these assumptions will be verified, one can propose for the myofibroblast a physiological role during development and in normal tissues where the production of mechanical tension is required, and a role in the evolution of normal and pathological wound healing as well as of fibrocontractive diseases. It has been suggested that during development connective tissue remodeling plays an important role in epithelial morphogenesis, implying a cross talk between epithelial and mesenchymal cells.[35] A pathological counterpart of this phenomenon could be the cross talk that starts to be understood between epithelial cancer cells and myofibroblasts of the stroma reaction.[37]

Conclusions and Perspective

The concept of myofibroblast has generated a significant amount of research during the last thirty years. It appears that rather than being a typical contractile cell, the myofibroblast plays a remodeling function that is necessary during development and repair phenomena. Many aspects of myofibroblast biology are not yet clear. We indicate arbitrarily here some of them that stimulate particularly our curiosity:

1. Very little has been done in the field of myofibroblast and of fibroblast heterogeneity, although early observations have shown that the agonists stimulating myofibroblast contraction are different for myofibroblasts derived from different organs.[38] Recent work has described markers distinguishing among different fibroblastic phenotypes.[39] Work along these lines would bring an important contribution to the understanding of fibroblast biology and function.

2. Myofibroblast apoptosis is a well-established phenomenon,[16] but its mechanisms are at present mysterious. Their understanding will help explaining the onset of pathological scarring and of fibrocontractive diseases.

3. A clear knowledge of the cellular origin of myofibroblasts in different pathological phenomena will be instrumental for the planification of therapeutic strategies.

4. Understanding of the mechanisms regulating force production and transmission to the extracellular matrix by the myofibroblast is essential in order to envisage their control. The recent progress in this direction points to a control of force production that is clearly different from that utilized by SMCs (for a review see ref. 32). A more precise knowledge of the control of stress fiber contractility is needed in this respect. The availability of the fusion peptide containing the sequence Ac-EEED[31] represents a useful new tool for the experimental work in this field and also a possible therapeutic agent for fibrocontractive diseases.

5. Little is known about the microenvironment allowing a cross talk between myofibroblasts and epithelial cells in different organs.[37] Further work in this direction will certainly contribute to the explanation of such important processes as organ morphogenesis and cancer evolution.

We are sure that during the next few years many of these questions will be answered and that some of the findings will represent the basis for new strategies aiming at the development of therapeutic tools for several important diseases.

Acknowledgements

This work was supported by the Swiss National Science Foundation, grant no. 31-68313.02. We thank Nature Reviews Molecular Cell Biology for allowing us to reproduce Figure 1 and Ms. E. Homberg for secretarial work.

References

1. Gabbiani G, Ryan G, Majno G. Presence of modified fibroblasts in granulation tissue and their possible role in wound contraction. Experientia 1971; 27:549-550.
2. Schurch W, Seemayer TA, Gabbiani G. Myofibroblast. In: Sternberg SS, ed. Histology for Pathologists. Philadelphia: Lippincott-Raven Publishers, 1997:129-165.
3. Desmouliere A, Darby IA, Gabbiani G. Normal and pathologic soft tissue remodeling: Role of the myofibroblast, with special emphasis on liver and kidney fibrosis. Lab Invest 2003; 83:1689-1707.
4. Desmouliere A, Guyot C, Gabbiani G. The stroma reaction myofibroblast: A key player in the control of tumor cell behavior. Int J Dev Biol 2004; 48:509-517.
5. Kissin EY, Korn JH. Fibrosis in scleroderma. Rheum Dis Clin North Am 2003; 29:351-369.
6. Redington AE. Fibrosis and airway remodelling. Clin Exp Allergy 2000; 30(Suppl 1):42-45.
7. Darby I, Skalli O, Gabbiani G. α-smooth muscle actin is transiently expressed by myofibroblasts during experimental wound healing. Lab Invest 1990; 63:21-29.
8. Gabbiani G. The myofibroblast in wound healing and fibrocontractive diseases. J Pathol 2003; 200:500-503.
9. Chiavegato A, Bochaton-Piallat ML, D'Amore E et al. Expression of myosin heavy chain isoforms in mammary epithelial cells and in myofibroblasts from different fibrotic settings during neoplasia. Virchows Arch 1995; 426:77-86.
10. Tomasek JJ, Gabbiani G, Hinz B et al. Myofibroblasts and mechano-regulation of connective tissue remodelling. Nat Rev Mol Cell Biol 2002; 3:349-363.
11. Border WA, Noble NA. Transforming growth factor beta in tissue fibrosis. N Engl J Med 1994; 331:1286-1292.
12. Desmouliere A, Geinoz A, Gabbiani F et al. Transforming growth factor-beta 1 induces alpha-smooth muscle actin expression in granulation tissue myofibroblasts and in quiescent and growing cultured fibroblasts. J Cell Biol 1993; 122:103-111.
13. Ronnov-Jessen L, Petersen OW. Induction of alpha-smooth muscle actin by transforming growth factor-beta 1 in quiescent human breast gland fibroblasts. Implications for myofibroblast generation in breast neoplasia. Lab Invest 1993; 68:696-707.
14. Grotendorst GR, Rahmanie H, Duncan MR. Combinatorial signaling pathways determine fibroblast proliferation and myofibroblast differentiation. FASEB J 2004; 18:469-479.
15. Serini G, Bochaton-Piallat ML, Ropraz P et al. The fibronectin domain ED-A is crucial for myofibroblastic phenotype induction by transforming growth factor-beta1. J Cell Biol 1998; 142:873-881.
16. Desmouliere A, Redard M, Darby I et al. Apoptosis mediates the decrease in cellularity during the transition between granulation tissue and scar. Am J Pathol 1995; 146:56-66.
17. Ross R, Everett NB, Tyler R. Wound healing and collagen formation. VI. The origin of the wound fibroblast studied in parabiosis. J Cell Biol 1970; 44:645-654.

18. Sappino AP, Schürch W, Gabbiani G. Differentiation repertoire of fibroblastic cells: Expression of cytoskeletal proteins as markers of phenotypic modulations. Lab Invest 1990; 63:144-161.
19. Kalluri R, Neilson EG. Epithelial-mesenchymal transition and its implications for fibrosis. J Clin Invest 2003; 112:1776-1784.
20. Yanez-Mo M, Lara-Pezzi E, Selgas R et al. Peritoneal dialysis and epithelial-to-mesenchymal transition of mesothelial cells. N Engl J Med 2003; 348:403-413.
21. Forbes SJ, Russo FP, Rey V et al. A significant proportion of myofibroblasts are of bone marrow origin in human liver fibrosis. Gastroenterology 2004; 126:955-963.
22. Hashimoto N, Jin H, Liu T et al. Bone marrow-derived progenitor cells in pulmonary fibrosis. J Clin Invest 2004; 113:243-252.
23. Abe R, Donnelly SC, Peng T et al. Peripheral blood fibrocytes: Differentiation pathway and migration to wound sites. J Immunol 2001; 166:7556-7562.
24. Yang L, Scott PG, Giuffre J et al. Peripheral blood fibrocytes from burn patients: Identification and quantification of fibrocytes in adherent cells cultured from peripheral blood mononuclear cells. Lab Invest 2002; 82:1183-1192.
25. Schmidt M, Sun G, Stacey MA et al. Identification of circulating fibrocytes as precursors of bronchial myofibroblasts in asthma. J Immunol 2003; 171:380-389.
26. Quan TE, Cowper S, Wu SP et al. Circulating fibrocytes: Collagen-secreting cells of the peripheral blood. Int J Biochem Cell Biol 2004; 36:598-606.
27. Serini G, Gabbiani G. Mechanisms of myofibroblast activity and phenotypic modulation. Exp Cell Res 1999; 250:273-283.
28. Hinz B, Celetta G, Tomasek JJ et al. Alpha-smooth muscle actin expression upregulates fibroblast contractile activity. Mol Biol Cell 2001; 12:2730-2741.
29. Chaponnier C, Goethals M, Janmey PA et al. The specific NH2-terminal sequence Ac-EEED of alpha-smooth muscle actin plays a role in polymerization in vitro and in vivo. J Cell Biol 1995; 130:887-895.
30. Derossi D, Joliot AH, Chassaing G et al. The third helix of the Antennapedia homeodomain translocates through biological membranes. J Biol Chem 1994; 269:10444-10450.
31. Hinz B, Gabbiani G, Chaponnier C. The NH2-terminal peptide of alpha-smooth muscle actin inhibits force generation by the myofibroblast in vitro and in vivo. J Cell Biol 2002; 157:657-663.
32. Hinz B, Gabbiani G. Mechanisms of force generation and transmission by myofibroblasts. Curr Opin Biotechnol 2003; 14:538-546.
33. Katoh K, Kano Y, Amano M et al. Rho-kinase—mediated contraction of isolated stress fibers. J Cell Biol 2001; 153:569-584.
34. Bogatkevich GS, Tourkina E, Abrams CS et al. Contractile activity and smooth muscle alpha-actin organization in thrombin-induced human lung myofibroblasts. Am J Physiol Lung Cell Mol Physiol 2003; 285:L334-343.
35. Doljanski F. The sculpturing role of fibroblast-like cells in morphogenesis. Perspectives in Biology and Medicine 2004; 47:339-356.
36. Schmitt-Graff A, Desmouliere A, Gabbiani G. Heterogeneity of myofibroblast phenotypic features: An example of fibroblastic cell plasticity. Virchows Arch 1994; 425:3-24.
37. De Wever O, Nguyen QD, Van Hoorde L et al. Tenascin-C and SF/HGF produced by myofibroblasts in vitro provide convergent pro-invasive signals to human colon cancer cells through RhoA and Rac. Faseb J 2004; 18:1016-1018.
38. Majno G, Gabbiani G, Hirschel BJ et al. Contraction of granulation tissue in vitro: Similarity to smooth muscle. Science 1971; 173:548-550.
39. Koumas L, Smith TJ, Feldon S et al. Thy-1 expression in human fibroblast subsets defines myofibroblastic or lipofibroblastic phenotypes. Am J Pathol 2003; 163:1291-1300.

Cytomechanics in Connective Tissue Repair and Engineering

Robert A. Brown*

Abstract

Mechanical forces are central to the control of 3D spatial organisation in connective tissue remodelling, repair and scarring. How this operates is increasingly seen as the next major research focus in this area. In contrast to mechanics at the tissue-scale, cell-level mechanics (or cytomechanics) is dominated by the cell-matrix-material interplay. The matrix completely modifies incoming, external mechanical cues whilst fibroblasts generate their own local forces to monitor and remodel that same matrix. By understanding these cell-material dynamics it is becoming clear how musculo-skeletal cells predictably adapt their responses to the perceived mechanical environment. Elements such as shape-change, orientation, matrix synthesis, migration can be explained relatively simply in terms of stress-shielding, matrix anisotropy, force vectors, rate of strain etc. In turn these translate into control of 3D tissue structure. (i) Anisotropy in fibrous collagen seems to be particularly important in modulating cytomechanical cues and so in controlling 3D remodelling. It is likely, then that the disorganised structure of scars resists subsequent remodelling because that same structural disorganisation obscures normal cytomechanical signaling. (ii) Reinterpretation of the role of cell force-generation suggests that this is necessary for (a) testing substrate material properties and (b) to produce incremental changes in tissue dimensions during remodelling. By understanding this interplay of cell-level forces with extracellular material properties (i.e., cytomechanics) new interpretations of connective tissue control and dysfunction are possible. Because of the predictable nature of (cyto) mechanics, these mechanisms are more direct and less complex than their biological counterparts.

Although mammalian connective tissues have a number of secondary functions (metabolic, depot, pseudo-regulatory), it is axiomatic that their PRIMARY role is structural and supportive. However, this mechanical function operates at both the macro and the cellular scales (micro-nano). Carriage of load at the macro, or tissue scale is familiar and relatively simple for us to understand in terms of the mechanical properties of the extra-cellular matrix (ECM) material. We are naturally familiar with the normal adaptation of structural tissues, their growth/ adaptation and reshaping to habitual use, aging and disease. We take for granted their repair and even fibrosis after injury and their ability to withstand the habitual mechanical loads placed on them. Our neuro-muscular systems are exquisitely tuned to use and to protect them. Yet to understand how these dynamics operate (and fail) we must look to the resident cells, which

*Robert A. Brown—University College London, Institute of Orthopaedics, UCL Tissue Repair and Engineering Centre, Royal National Orthopaedic Hospital Campus, Stanmore, London, HA7 4LP, U.K. Email: rehkrab@ucl.ac.uk

Tissue Repair, Contraction and the Myofibroblast,
edited by Christine Chaponnier, Alexis Desmoulière and Giulio Gabbiani.
©2006 Landes Bioscience and Springer Science+Business Media.

create and manipulate the ECM mechanical properties. This involves understanding the mechanical environment down at the cellular level; a much more tricky task. This world of cytomechanics is very unfamiliar and holds many surprises for the unwary biologist. Two examples, illustrating this topsy-turvy world are (i) the very limited DIRECT effects of gravity on cells, yet (ii) their exaggerated sensitivity to fluid viscosity.[1,2]

It is assumed as a fixed point here that understanding the mechanisms of growth, adaptation and restoration of ECM material properties must be the focus of research in our field. Control of 'material properties' (understood rather better outside the biological disciplines) is dominated at least as much by micro-architecture and polymer distribution as by composition and density. The assumption of this author is that it is essential to identify the PRIMARY environmental cues used by tissue fibroblasts and the way in which they might operate, or fail. Since fibroblasts live in a sea of load-bearing polymer aggregate with the role of maintaining mechanical function, it seems inevitable that some of the most potent PRIMARY environmental cues will be mechanical. This chapter explores current concepts and knowledge of how cells monitor, respond to (and potentially misinterpret) local, micro-scale mechanical forces.

Stress-Shielding of Resident Cells against Applied Forces

More than a century post-Wolff,[3] interpretation of cell function in terms of mechanics and material properties is natural for hard-tissues, but surprisingly restricted for adult soft tissues. At the human scale we have first hand experience of the dominant forces applied to connective tissues; tension in the soft tissues, compression in the cartilages and bones and shear in flow or motor tissues such as muscles and the vascular system. However, even these simple generalisations tend to unravel, particularly at the cell level, for two main reasons. Firstly, whatever the mechanical loads applied to biological ECMs (i.e., relatively compliant/deformable) they will tend to generate complex (sometimes localised) mixtures of tension, compression and shear in the materials. For example tension applied to a ligament will produce compression at the core and shear between fibre bundles as it bends. Secondly, the physiological loads which pass through most musculo-skeletal tissues would inevitably be lethal if applied directly to the resident fibroblasts. Rather, connective tissue cells elaborate

---

Box 1.

Modern bio-molecular concepts often place great importance on growth factor/cytokine control of tissue function. However, it is increasingly clear that in mechanical tissues, such as connective and muscle tissues mechanical forces will be disproportionately important as primary environmental information. For example, whilst diffusible factors are effective in controlling the rate processes, their ability to *control* 3D spatial organisation (i.e., tissue architecture) in adult tissues is likely to be very limited. Spatial control via soluble molecular signalling would depend on diffusion gradients, as in embryonic development. However, signalling by macro-molecular gradients of more than a few microns through dense, highly motile (i.e., mixed) adult tissues would become impossibly slow, complex and unstable. In contrast, we know from engineering that mechanical signalling is excellent for long-range monitoring/transmission of spatial and material information. Mechanical cues:
- Can be propagated over long distances (up to metres),
- Always have one or more vector (directional element) and so carry spatial information,
- Comprise a combination of tension, compression and shear (one normally being principle),
- Carry information in their magnitude, rate of change and/or frequency
- Are modified by the solid/fluid medium to give interpretable material properties information.

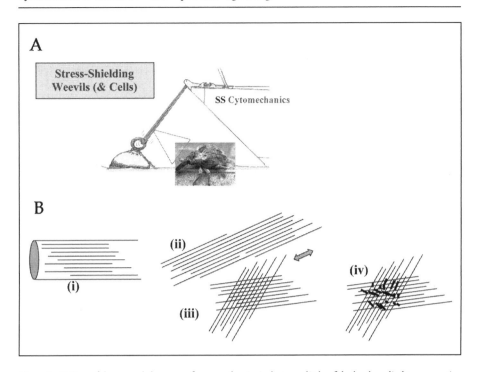

Figure 1. A) One of the central elements of cytomechanics is that very little of the load applied to connective tissues can act directly on resident cells (even a small fraction would be lethal). Rather, fibroblasts etc. are stress-shielded by their collagen ECM. This can be illustrated by considering a boat moored through a rope. The rope material properties (strength, elasticity etc.) determine how well the boat rides at anchor against tides and winds (analogous to tissue-level mechanics). However, the rope weevil, living in that rope, is not the slightest bothered by wind, tide or rope strength (analogous to cell-level, or cytomechanics). In fact the weevil is almost completely stress shielded, much as a tenocytye between its collagen fibre bundles. Each carries on its business between the strands, which carry the external loads. The KEY difference is that the weevil has no ability to alter the rope whereas fibroblasts are continually monitoring and modifying the material properties of their ECM—apparently to keep pace with changes in prevailing loads. This, then, describes the starting hypothesis. B) A second important point is that for almost all connective tissues in which we are interested, the load-bearing function is completely dependent on collagen fibril architecture. Clearly in the rope example, the same mass of hemp assembled as a nonwoven 'wool' would have little ability to tether the boat. Similarly, almost all native collagen networks are organised and aligned (i.e., anisotropic fibre orientation), varying from almost parallel in ligament/tendon (i) to multi-layered, as in dermis. Groups of fibres/fibrebundles can be parallel to each other (ii), lie freely at an angle in different tissue layers (i.e., in the z-plane, allowing sliding and angle change between layers (iii)) or be linked in the z-plane to transmit loads through the depth of the tissue (iv). Clearly these are adaptations to carry anything from near uniaxial tensions to dissipation of complex dynamic loads through all planes.

an ECM in which the polymer fibres, architecture and/or interstitial fluid carries the vast majority of the working tissue loads. Cells, then, are stress-shielded by the matrix, which they make,[4] such that they are exposed only to minor deformations (illustrated in Fig. 1). Not just production but also adaptation of this 'stress-shielding ECM' appears to be a major part of the normal fibroblast behavioural programme, as we shall see. The inevitable consequence of heavy stress-shielding of cells is that those limited deformations (strains) which they do undergo are complex and largely dictated by matrix architecture (which, of course, they adapt). This analysis is a key element in cytomechanics.[5]

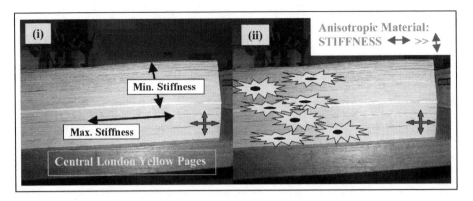

Figure 2. The importance of anisotropy in fibrous cell substrates can best be illustrated by reference to the London yellow pages. It is easy to understand that this surface is highly anisotropic. It is very stiff in the long axis and compliant (i.e., the pages flick) in the perpendicular, short axis. If fibroblasts could be grown in the yellow pages they would 'feel' great resistance when they applied forces in the long axis (i.e., the substrate would stay still and they would move) but a rather 'soft' substrate in the short axis (i.e., the substrate deforms and the cells stay put). Cells generating tensile forces randomly in all directions on this substrate would tend to elongate and move mainly in the long (stiff) axis. Equally, external forces applied in the long axis would have little affect on attached cells but those in the compliant, short axis will severely deform, and so 'activate' them (see Fig. 9 and the OOPS principle: Box 5).

Force Vectors and ECM Anisotropy

Unlike (bio)chemical cues, mechanical forces have 'direction', or vectors. Indeed, this additional information underpins the importance of mechanical forces in controlling 3D spatial organisation and so tissue architecture. The operation and interaction of co-acting force vectors (from space flight to atomic collision) are well understood and relatively predictable. Examples of both the propagation and dissipation of external force vectors are common in human connective tissues (illustrated in Fig. 2). For example, collagen organisation in ligaments and tendons is adapted to faithfully transmit dominant force vectors whilst intervertebral disc dissipates vertical spinal loading as widely as possible.[6] Importantly, these two examples also illustrate the strict collagen fibre anisotropy maintained and adapted by fibroblasts. Whilst it is possible to understand the role and mechanism of action of these highly ordered collagen alignments and architectures—in terms of tissue mechanics[6-8]—it is far less clear which mechanical cues are important at the cell level to maintain and even adapt them in the face of changing habitual loading. In a highly simplified 3D model system (under uniaxial load) it was possible to demonstrate a relatively simple relationship between applied load and cell elongation and orientation. In this case fibroblasts align parallel with the principal strain through that matrix location (Fig. 3). However, where there is no dominant aligning strain (multi-vectored) cells take on a stellate, nonaligned morphplogy.[9,10]

The Central Problem of Scarring Is Collagen Matrix Contracture

Scarring of collagenous tissues is a central problem in repair biology, particular after extensive loss of tissue mass (Fig. 4A). This is true of almost all vascular connective tissues, from skin and tendon to nerve, fascia-vessel wall and muscle. In some cases the 'scarring ' is characterised by an excess of collagen fibrous tissues (e.g., in lung, heart, dermal keloid). However, there is almost always also a substantial loss of collagen fibrillar architecture, as seen in dermis, cornea, and ligament etc.[11,12] Interestingly, this inability of connective tissues (with a few exceptions) to spontaneously (re)construct collagen 3D organisation is now recognised as a central problem in tissue engineering.[4,13] Significantly, the most effective forms of adult tissue regeneration and of successful tissue engineering involve the reassembly of epithelial tissues such as

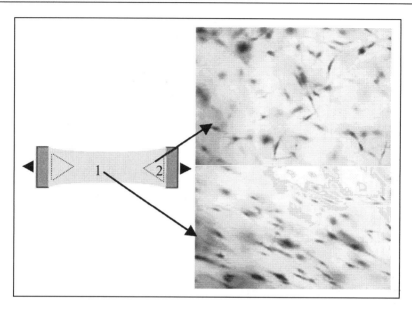

Figure 3. In a 3D rectangular collagen gel (FPCL) tethered uniaxially (hence generating tension chiefly in its long axis) finite element analysis of the dominant strains suggests that there are 2 main, contrasting regions. The bulk of the gel is under substantial strain (deformation) in the long axis. However, a triangular ('delta') zone with multiple strain vectors is created at each end by the presence of the stiff (hatched) anchoring bars. Cell morphology and alignment are a reflection of this mechanical environment. Under loading, fibroblasts in the bulk of the gel become bipolar and parallel to the principal strain. Those in the delta zones are stellate and nondirectional.[9]

keratinocytes in the epidermis.[14,15] It is notable that in such cases much of the spatial organisation is based on elaborate cell-cell self-assembly producing sheets and layers with little 3D organisation. Poor reassembly of 3D collagenous structures is thought to be largely due to ECM contracture *during* the repair process,[16] which set up random tissues tensions and force vectors (Fig. 4A). This, then, is the rationale to understand the mechanisms of contracture and cytomechanics in 3D spatial control, as it relates to scarring and tissue engineering.

Growth Repair and Contracture of the ECM: What Do They Involve?

From the stage of birth onwards, soft connective tissues of the body (tendon, ligament, skin, facia etc.), all increase many fold in size and volume. This is primarily by deposition of new collagen-rich ECM with a gradual proportional decrease in the cell content. Through the process, tissue dimensions increase by additional of new ECM throughout the entire tissue[17] (i.e., not at the ends or edges). Figure 4B, shows how stapling and x-raying a growing rabbit ligament over 6 months where the length increased. In this case every part of the ligament became longer, consistent with an interstitial growth process. This is a key example tissue as we can be sure it was under tension and carrying functional load throughout the period (i.e., there was no unloaded opportunity for the structure to be degraded and fully reassembled).

Importantly, there is no real engineering equivalent to help us to understand this process. It seems to involve each cell laying down new matrix within the structure. Clearly the result of this is that all the surrounding cells must reorganise (remodel) their own matrix to accommodate the new material. Whilst there is currently little clear understanding of how this is orchestrated at the cell level or even how the collagen network remodels, it is thought that it involves controlled slip between fibrils.[1,5,18,19] It seems reasonable though that it is a similar cell/molecular process for both ECM growth and contracture (i.e., ungrowth) under opposite mechanical drivers.[5]

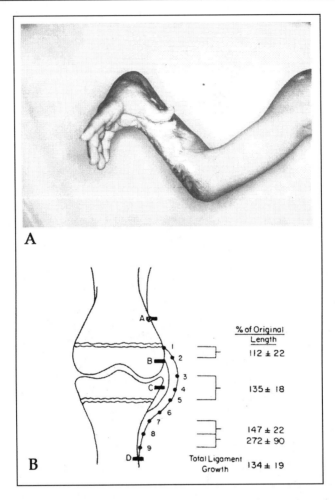

Figure 4. A) Flexion deformity caused by a burn scar 'contracture' over the wrist. Note the hand locked into a flexed position despite the relatively massive 'contractile' forces generated by forearm skeletal muscles (photo kindly provided by Dr. Jaysheela Mudera). B) When the relative positions of metal markers within the collateral ligaments of a growing animal are plotted over its main growth period it was clear that this tissue enlarged throughout its complete substance.[17] This represents unambiguous evidence for 'interstitial' growth of a soft connective tissue, which is both under constant functional load and rigidly anchored throughout the growth period (from: Muller P, Dahners LE. A study of ligamentous growth. Clinc Orthop Rel Res 1988; 229:274-277).[17]

Box 2.

For the purposes of this review the terms 'contraction' and 'contracture' of collagen lattices will be rigorously distinguished. Contraction is taken to be a reversible, energy dependent, muscle-like process of compaction, maintained by cell-generated force. Contracture is the altering of the collagen lattice material properties and architecture to produce a geometric reduction in material length by physical fixing of the compacted structure (Fig. 4A).

Tissue remodelling is central to wound repair in most connective tissues. Research into its control mechanisms has the clinical aim of learning how to influence the restoration of functional tissue architecture. This is against the background that scar tissue, which is the default outcome of tissue repair, is poorly ordered with disrupted architecture.[11,12] Consequently, a clear understanding of the mechanisms of 3D spatial remodelling seems to be a prerequisite for progress. By this analysis, Collagen 'remodelling' will be involved both in the GROWTH of new tissue to fill a defect and shrinkage during contracture.[16] Collagen contracture and the myofibroblast are widely thought to be important in both for disruption of 3D tissue architecture and the poor scar ECM material properties.[16,20] Let us then examine the process and meaning of force generation by (myo)fibroblasts in collagen matrices.

Tissue Tension and Cell Force Generation

In Cytomechanics and ECM function Newton's 3rd Law of motion is particularly important (Action and Reaction are equal and opposite). Forces generated locally affect all attached structures and many of these are soft (i.e., compliant/deformable). As a result, it is not always clear (for example by fixed image microscopy) where tensions are generated as opposed to being propagated. One unambiguous example of tissue tension has been described in an animal mode for joint laxity,[21] as it involves changes in ligament dimensions between 2 fixed bony attachments. In short, one end of a trans-joint ligament attachment is surgically detached and shifted (such that the ligament becomes loose (Fig. 5). However, despite induction of a 3·3 mm laxity the joint was found to return to normal tension after only 3 weeks (through a

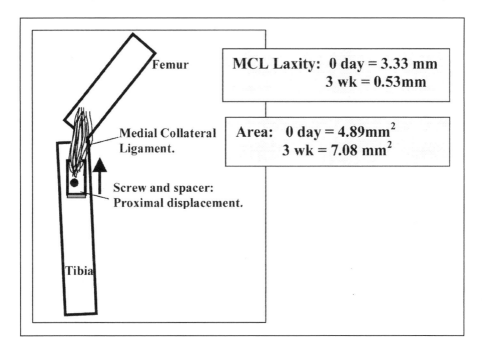

Figure 5. An experimental model from Wallace et al[22] demonstrates that connective tissues not only have an inherent tension (due to their anatomical length) but that this tension is actively monitored and maintained. In this model, rabbit knee collateral ligaments were surgically de-tensioned by repositioning their bony attachments, as shown. Without further treatment, the resulting joint laxity of 3.3 mm was eliminated in only 3 weeks (with compensating increase in cross-section). In a functional, anchored, load-bearing tissue such as this the only current explanation is that the material of the ligament had been physically shortened by its resident cells (adapted from Wallace et al).[22]

decrease in length and corresponding increasing in cross-section). The body of the ligament itself was not injured. This key model teaches us a number of lessons:

- Remodelling to a shorter ligament must represent physical shortening of the collagen network (it is inconceivable on energy grounds and others, that the few resident fibroblasts could maintain a 'contraction' against such loads).
- The process can be remarkably rapid and triggered by changed mechanics (rather than injury/repair).
- It is a normal process and again does not involve widespread tissue destruction/regrowth.

This example, then, illustrates the enigma we need to explain. Clearly it must involve some form of internal force generation in which the (myo) fibroblast lineage is likely to play a role.[22] But according to this analysis, force generation is not the end product but rather part of the mechanism by which the collagen scaffold is shortened (i.e., remodelled).

In the following sections possible mechanisms and mechanical controls of this process will be explored using the 3D fibroblast populated collagen lattice (FPCL) models.[23-41]

Cell (Fibroblast) Generated Force: What Is It and What Is It For?

Cells of the fibroblastic lineage (i.e., dermal, tendon, fascia etc.) generate contractile forces through the conventional actin-myosin motor, controlled via a complex array of intracellular signalling pathways (beyond the scope of this chapter, see ref. 5). An important, if self-evident feature of ECM contraction is that ECM contraction can only occur where this cell-motor apparatus is mechanically attached to the matrix. In the 3D collagen model (FPCL) there are really only two plausible mechanisms by which cell-generated force can be transferred onto the substrate, and so appear as externally measurable. These are (i) direct cell-ECM, integrin-mediated linkage,[23,24] or (ii) indirectly by cell-cell attachment with entrapment and deformation of intervening collagen lattice (Fig. 6). Cell-ECM linkage appears to dominate in fibroblasts, as judged by the effects of experimental manipulation of ligands and integrin availabilities.[23,24,38] Where force is measured as the direct output (using the Culture Force Monitor model)[26-32,41] it was possible to demonstrate not only the involvement of integrin-mediated cell attachment via fibronectin, vitronectin and collagen, but also that there appears to be a sequence to this attachment.[24]

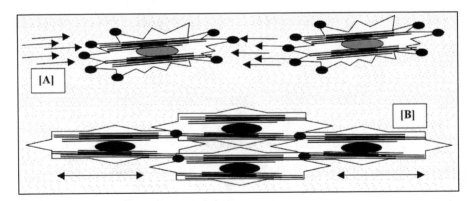

Figure 6. Two models for cell-force generation against their substrate attachment. Model (A) (upper, fibroblastic model) suggests that cells attach directly to fibrils of the collagen matrix (attachments indicated as black spots). When these cells apply tensile forces to the attachment points the collagen lattice itself deforms locally. In contrast, it is likely that some cell types with greater tendencies to form cell-cell attachments, notably skeletal myoblasts (B), will deform the collagen lattice indirectly as a consequence of the contraction of groups of cells.

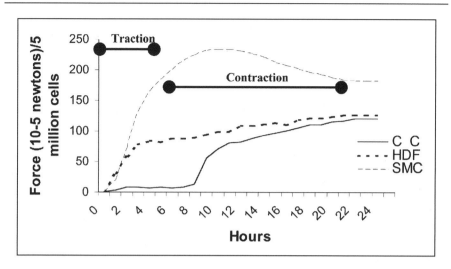

Figure 7. Force generation by three cell types, smooth muscle cells (SMC), human dermal fibrobasts (HDF) and C₂C₁₂ skeletal myoblasts(C C) (adapted from Cheema et al[39]) over a 24-h time period. The distinctive shape of the first contraction periods (0 to 8 h) are characteristic for each cell type and largely due to traction as the cells move and extend processes through the collagen lattice. This gives way to more conventional cytoskeletal contraction against the matrix.

Type (ii) transfer of cell-force (indirectly) to the ECM becomes a possibility where the seeded cells have a strong tendency to form cell-cell cytoskeletal contacts. This has been identified in collagen lattices seeded with vascular endothelial cells[30] or with skeletal myoblasts.[39,40] In the latter case, high-density myoblasts begin to form cell-cell contacts as their proximity increases and this corresponds with a gradual generation of measurable force output in the CFM model.[39] Figure 7 illustrates this effect in terms of the different force-time profiles between fibroblasts, smooth muscle cells and skeletal myoblasts, over the same period. Smooth muscle cells not surprisingly, generate around 2-3 times more total force than dermal fibroblasts, though with strikingly similar initial gradients. It is not yet certain how much cell-cell linkage contributes to the force generation in smooth muscle cells (particularly its peak level). However, it is likely that almost all of fibroblast generated tension is through cell-collagen attachment.[23,24] In contrast, skeletal myoblasts demonstrate a lag-period before force generation, correlating with spreading of cell processes. The formation of cell-cell interactions is a characteristic of single myoblasts in the prelude to fusion of myotubes.[42] Clearly, this phase will be dominated by the availability of surface receptors (cell or matrix) for both types (i) and (ii) collagen contraction.

It is suggested that force generation in this period is chiefly due to 'traction' or cell locomotion across/through its substrate,[5,36] related mainly to cell motility. In the FPCL model this phase is artificially synchronous, as all cells start (time 0, Fig. 7) as round and non-attached.[43,44] In vivo parallel can be drawn with forces generated as waves of fibroblastic cells migrate into the repair site from marginal tissues,[45] or with morphogenic tension during embryonic development.[36] In contrast, the relatively constant force output once cells have spread (Fig. 7) seems to be a feature of less mobile cells generating a 'contraction', with a progressively more extensive stress fibre network. It has been suggested that this reversible cytoskeletal maturation (the 'protomyofibroblast' stage) occurs in reaction to the increasing stiffness of the collagen substrate.[5,25] This common pattern seems likely to reflect fibroblast mechanoresponses during rapid formation and reorganisation of soft granulation and repair tissues.

Whilst TGF-β1 is now closely associated with α-smooth muscle actin expression (and by implication myofibroblast differentiation and increased force generation) this stimulation characteristically takes a number of days.[46,47] Direct stimulation of TGF-β1 on force generation has been shown,[35] but this occurs either immediately, during the traction phase (in serum depleted)[28] or in the contraction phase (with serum) (Marenzana et al submitted). This suggests that TGF-β1 acts (with serum elements) to stimulate cell contractile and attachment mechanisms, dramatically increasing force output. In turn, it is reasonable to expect that prolonged cell tension applied to a stiff substrate would promote myofibroblast differentiation.[5,48]

What Then Are the Real Functions of Fibroblast Force Generation?

There is clearly a mechano-feedback response between groups of cells as they produce an increasingly dense and stiff ECM. For example, fibroblasts in their contraction phase (Fig. 7) alter their force generation (up or down) to counteract small changes in external applied load. This is consistent with a collective tensional homeostasis.[32] The concept of a cell-level tensional homeostasis is key to cytomechanics, and not difficult to envisage as an inbuilt response of the cytoskeletal motor, to stretch or relaxation. Its existence supports the idea that cells maintain a small applied load on their substrate as a means of monitoring ECM material properties (e.g., stiffness, elasticity, anisotropy).[49] This is clearly an essential fibroblast role as they lay down and remodel ECM to resist local, repetitive loads, for example in wound repair. Indeed, measurement of substrate deformation in response to known applied loads is essentially how materials scientists determine materials properties. The reader is encouraged to examine how we distinguish (blind) between sheets of glass, plastic, fabric and rubber by feeling how each one deforms in response to small known tensions, applied manually. However, it is then essential to distinguish between deformation due to the applied as opposed to any external loads (hence the need for a tensional homeostasis).

A second function of cell-generated tensions is for compaction of initially loose, newly deposited collagen into a dense aligned (anisotropic) matrix, capable of carrying muscle and gravitational loads. The packing of collagen into a denser material (with larger diameter fibrils) has been identified in the FPCL model, as 'bundling'.[50,51] This process almost certainly occurs as new ECM undergoes dehydration by fibre packing and fluid expulsion. When resident fibroblasts apply loads directly to collagen fibrils the fibrils will move towards the cell at the same time that the cell moves over the fibril (in proportion to the matrix stiffness: Newton's 3rd law). This can be extrapolated in 2 directions.

- If that collagen is 'fixed' into its new position (i.e., remodelled into a new material geometry) then a constant cell generated force will gradually increase the local ECM stiffness. In time increased substrate stiffness would allow the cell to move more, and so leave that area of substrate. On reaching another area of lower stiffness (fibril density) its locomotion will again be slowed down until that area of substrate increased in stiffness to a level supporting cell locomotion (Fig. 8). Such a repetitive, substrate-stiffness driven system (compaction-migration) would continue for as long as cells generated sufficient tension and areas of low density, low stiffness collagen became available. This then represents a hypothesis for near automatic material remodelling hypothesis, based on the predictable behaviours of (fibrous lattice) materials and mechanical loads. It is a form of the ratchet hypothesis (built on Newtonian mechanics) for collagen contracture[5] explaining a natural effect, which we know, occurs.

- However, if compaction of the loose collagen substrate is not fixed, then both the shortening of the matrix and the increase in stiffness are conditional on the cell force generation. That is, the new (remodelled) material properties and geometry will return to baseline, once the cell-tension is removed. This case more closely resembles the wrinkling of a silicone membrane under (myo)fibroblast contraction. Removal of the cell-force abolishes membrane deformation with little accompanying cell locomotion.[52]

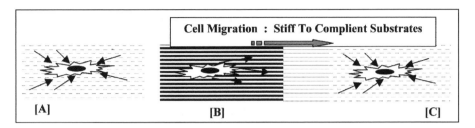

Figure 8. Illustration of the hypothesis linking cell-force generation, compaction of compliant substrates and cell motility. This hypothesis explains how the process of collagen lattice compaction and increasing stiffness could be a simple, automatic consequence of cytomechanics. Any strongly contractile, motile fibroblast within a loose (nonstiff) collagen lattice (A) will pull together the surrounding fibrils (potentially into an alignment under uniaxial tethering) (B). As lattice stiffness (and anisotropy) increases the collagen will move less and the cell will move more (B), i.e., the cell migrates away (potentially along the anisotropy, (Fig. 2). Migration will slow down again when the cell enters another soft or compliant region of matrix (C), and so the process continues.

These two mechanisms (fixed and unfixed matrix remodelling) may well operate in the same ECM, depending on a cell's ability to (a) bond collagen fibrils or (b) to generate sufficient tension. This, then, identifies a central focus for cytomechanics, towards understanding how/when fibroblasts can 'fix' geometric changes in collagen fibril packing: more of this later.

Cell-Molecular Responses to Cytomechanical Cues

Many studies have now established that simple changes in the forces across cells (i.e., the mechanical environment) have major effects on the on cell behaviour, gene and protein expression.[27,38,40,53,54] Examples include stretch in skeletal myoblasts (up regulation of the smpx gene,[53] tension versus nontension in fibroblast seeded collagen lattices,[54] collagen synthesis on cyclically stretched silicon membranes[55] and changes in proteoglycans synthesis in compressed chondrocytes.[56,57] Studies on single cell mechanics can provide useful clues[58] but often cannot include the critical role of the matrix. On the other hand, complex 3D multi-cell systems ('tissue models') are only slowly giving useful analyses, as the basic material-cell properties are understood[32,34] (e.g., contribution of cells or loading frequency to overall stiffness). For the most part it remains difficult to determine in any detail what loading regimes operate at the cell level and studies continue to show that altered mechanics alter cell responses, but provide little further insight.

    i. Perhaps one of the most persistent difficulties is in the concept of a mechanical (nul) control for cultured cells. Although cells in monolayer are frequently taken to be 'unstimulated' they are actually under considerable (variable) reactive tension due to the force they themselves generate against the noncompliant plastic substrate.[59] Indeed this isometric tension may well help to drive fibroblasts to proto-myofibroblasts (stress-fibre rich) and α-smooth muscle actin positive myofibroblasts.[5] This relatively static, endogenous stress is just as much a loaded environment as an externally applied cycle, with vectors and resultant strains on adjacent cells. In fact, almost no cultured cells are under nil force (just as cell cultures are never 'protein-free'). It is essential, then, for sound cytomechanical models, that the principal force types, vectors, frequencies and rates are defined.

    ii. The second basic controversy is between so called 2D and 3D culture systems. It is frequently argued that '2D' cultures are slightly simpler (arguably less physiological) models than 3D. Cells in monolayer obviously have a ventral (attached) dorsal (nonattached) polarity, i.e., a clear directional cue. However, from the field of micro and nano-topographical control of cells it is increasingly clear that the scale at which cells take surface cues from their matrix, we should reconsider this idea of '2D'. From early clues that cells react to

Box 3.

Fibre anisotropy (acquisition of a predominant alignment) is a key part of tissue architecture and collagen is the main element in this. Tissue anisotropy can be induced by the presence of a preexisting substrate alignment (or a plane of stiffness), which guides cell alignment, movement and so matrix deposition.[10,59,61] However, in many cases tissue anisotropy seems to be produced/maintained by predominant and repetitive alignment of strain through the structure. This can be by external forces, by virtue of preexisting muscle and skeletal anatomy (e.g., predominantly uniaxial loading along the digits). Endogenous, cell-generated tensions acting between asymmetric or uniaxially opposed anchoring points can realign existing fibres and cells[9] and direct the deposition of new collagen (Marenzana M, Brown RA et al, submitted). An important feature of large-scale injuries is the loss of many such local anatomical features or anchor points, with consequent degradation of spatial mechanical cues available to new repair tissue and scars.

ultra-fine detail at the scale of collagen fibril surfaces[60] it is now clear that cytoskeletal responses are triggered by surface features down to 13 nm.[61,62] Interestingly, this work is now generating concepts, which converge with those from cytomechanics on how cell-generated forces are used to monitor topographical and material properties of the substrate.[59,63] From the cell perspective, then, so-called 2D cultures become 3D cultures where the smallest (i.e., vertical) dimension is completely uncontrolled. At this nanometre scale, grooves, pits, absorbed proteins, matrix aggregates and certainly other cells can constitute significant surface features, presenting a variety of (normally uncontrolled) cues, e.g., mechanical stiffness, dimensions, orientation. Similarly, cultures, which seem to be '3D' at the millimetre scale (e.g., meshes or sponges >100 μm pore size) actually present a series of textured but predominantly planar surfaces at the cell scale.

Despite these constraints increasing numbers of sound cytomechanical systems are now being used to determine the mechanisms by which the cell-scale mechanical environment controls cell molecular responses.[64]

Collagen synthesis by dermal fibroblasts was found to increase substantially with cyclically loading, particularly over longer.[55] Interestingly this was related both to increased synthesis and an increase in collagen deposition, due to enhanced procollagen peptidase activity and so an increased rate of collagen processing.[65] In 3D collagen gels, a 2 to 8 fold increase in collagen synthesis and incorporation was seen under uniaxial, low frequency loading within 24 h at 1 cycle per hour (Parsons M. 2000, PhD thesis, University of London). Studies on tissue engineered 'tendon' templates have identified comparable increases in collagen deposition and mechanical properties over 3D culture periods from 1–12 weeks. Work from the Banes Laboratory[66] using cyclic loading (1% strain, 1Hz for 1h/day) reported protein collagen remodelling with an increase in ultimate tensile strength from negligible to 0·33 MPa. in 7 days. Other approaches have highlighted the biological importance of very low rates of strain (i.e., cyclical loading frequency) on fibroblast-collagen systems, particularly in highly compliant, immature tissues (and their models).[9,32,67] These studies begin to define the frequency, type and dominant force vector which control matrix remodelling at the cell level.

Early studies by Prajapati et al[26,69] on matrix metalloproteinase production (MMP: enzyme activity level)) found complex responses in 3D fibroblast populated collagen gels loaded under defined mechanical conditions. These again found that fibroblasts in the highly compliant (collagen gel) scaffolds are sensitive to low strain/low frequency uniaxial loading. These experiments also demonstrated that force vectors and strain pattern across cells influences MMP activity. MMP 1, 2 and 3 gene expression was used in a modified 3D FPCL system, to test the importance of cell alignment and force vector.[10] In this model, cells were held in an elongate morphology, perpendicular (rather than parallel to) the applied strain and their response compared to cells, which were free to take on the normal parallel alignment. This forced perpendicular alignment

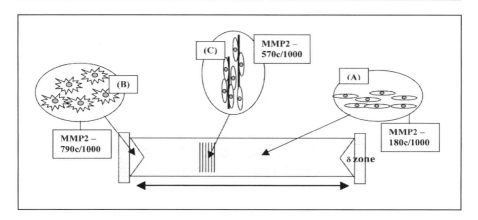

Figure 9. Diagram illustrating the analysis of opposing dual cues on fibroblast matrix metalloproteinase (MMP-2) gene expression (adapted from Mudera et al[10]). MMP-2 expression was measured (copies per 1000 cells: c/1000) for the nonaligned delta (B) and aligned (A) zones after 16 h uniaxial cyclic loading (double headed arrow). This was compared with expression by cells in a zone (C), which were held elongate but perpendicular to the applied strain. The dramatic fall in MMP-2 expression associated with the ability of cells to align with the principal strain was largely abolished in the dual cue (out of plane) zone, (C). This effect was seen even though around 80% of the cells in this zone were not directly attached to the contact guidance strands. Since the realigned parallel with the strain, as in zone (A), the extent of stimulation is likely to be underestimated.

was achieved by insertion of contact guidance fibres across the loaded axis of the FPCL (Fig. 9). Parallel-aligned cells (A) showed a four-fold drop in MMP expression relative to those in the nonaligned gel zone -B- (the delta zone). However, this down-regulation was abolished in the zone where cells were held perpendicular to the principal matrix strain (C). This highlights the importance local force vectors in the cell response to mechanical forces.

Further insight into gene response to tensile loading has come from a similar 3D collagen gel model of skeletal muscle, based on the FPCL.[39,40] Application of uniaxial loads either as low strain-rate ramps or cyclical loading (between 1 and 10 cycles/h) induced changes in two splice variants of the growth factor, IGF-1 (IGF-1Ea and IGF-1Eb: or MGF[70]). Importantly, both isoforms exhibited distinct strain-dependent responses to loading, though in differing directions. At critical strain-rates, expression of the mechano-growth factor (MGF/IGF-1Eb) increased by 2 to 3 orders of magnitude. These very different movements between splice variants of the same gene suggest the possibility that cell responses are to changes in isoform RATIO as well as overall levels.

Box 4.

The frequencies selected for experimental cyclic loading of cultures can prove difficult. Whereas a few (adult) tissues may be subject to frequencies around 1 Hz., this is probably not true of the cells we aim to model, i.e., fibroblasts in soft, repair matrices. Certainly the rate of change of strain under endogenous cell force generation in 3D culture is low.[32,40] In particular, the standard 3D collagen gel substrate (like most biological tissues) changes from elastic to viscoelastic in behaviour,[68] as the water displacement becomes limiting. Importantly, these experimental gels behave elastically only up to ~10 cycle/h (or 0.003 Hz) at standard strain (Marenzana, Brown, Cheema, in preparation). At greater strain rates (frequencies) the increasingly viscoelastic gel becomes stiffer and so deforms less than predicted (i.e., at the cell level, another form of shielding).

Contraction and Contracture for Collagen Remodelling (Shortening)

As discussed above, a key factor in disruption of repair tissue architecture and mechanical function to be (myo) fibroblast-mediated contracture.[16] It is important then to understand what these cells are mediating. Glimcher and others[5,71] have pointed out that contracture of connective tissues cannot be the high energy, reversible process seen in muscle tissues. Indeed, whilst cell force generation must be important, contracture is in fact a process of ECM material shortening (remodelling) rather than a muscle-like contraction. Hence the processes of contracture,[71] cell independent contraction[35] and tissue tension[21] are in fact consequences of fibroblasts altering the dimensions of their collagen substrate.[5] Obviously, altering the dimensions of anchored connective tissues must alter tissue tensions (Newton's 3rd Law: Fig. 5). It is likely that the number of pathologies we recognise as associated (presently dermal scar (Fig. 4) and Dupuytren's contracture) will increase.

Tomasek et al[5] recently proposed how this normal fibroblasts/myofibroblast process could act to maintain and adapt both the material properties (strength, stiffness etc) and geometry of the ECM, through cell and matrix tension. We have now identified this process in vitro (reengineering of a 3D collagen lattice by tendon fibroblasts), in terms of increased fixed tension (matrix shortening) and matrix strength[72] (and Marenzana et al submitted). This is measured as an increase in fixed matrix tension after complete removal of cell-generated force (i.e., collagen lattice was bonded shorter and so changed as a material). It is now clear that this is a very real, tension—and time-dependant process, influenced by environmental factors, TGF-β and externally applied loading. However, these environmental factors also critically controlled matrix material properties, apparently by differences in bonding strength.

Consequently, we can now manipulate and measure this critical material remodelling process. In turn, this teaches us how (myo) fibroblasts use cytomechanics, local tensions and the fibrillar anisotropies of collagen to adapt the dimensions and material properties of their ECM. Clearly, we also then begin to understand how and when the process fails.

Summary and Conclusions

The central cell processes involved in connective tissue remodelling are heavily regulated by mechanical forces. This is particularly true for control of 3D spatial organisation and remodelling, which is the key fault-line in repair biology and scarring research. However, the scale of

---

Box 5.

A frequent question in cytomechanics is what special types of mechanical force are able to bring about cell responses in vivo, given that our tissues are under constant muscle and gravitational load. If cells can be activated in culture why are resident cells of our connective tissues not in a continual state of activation? Clearly, we must look for clues amongst the unusual or aphysiological facets of everyday mechanical loading. The most obvious candidate, simple total force magnitude, is probably the least likely, except in extremis. Doubtless, extreme loads can stimulate cells but these would need to be great enough to cause permanent matrix damage. Oddly, there is circumstantial evidence that complete removal of tissue tension may be a plausible candidate. Orthopaedics is filled with reports that tendon stumps become too weak to suture only for a few days after rupture. Cell and matrix changes are evident in explanted tendons 72 hours after explanting and tenocytes will migrate out of ex-vivo tendon cultured within a collagen gel over the same time scale (Hodgkins, and Brown, unpublished data). However, the 'out-of-plane' study described above[10] also suggests a clear candidate on cytomechanical grounds. Fibroblast elongation and alignment parallel to a habitual principle strain[9] suggests that they use shape to minimise strain (example illustrated in Fig. 10). This is consistent with the finding that elongate cells trapped perpendicular to an applied strain are maximally stimulated.[10] Hence, force vector, relative to matrix and cell orientation, represents a key to cell stimulation. We have termed this 'Out-of-Plane-Stimulation', or OOPS.

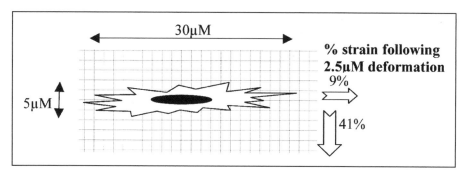

Figure 10. Example to illustrate effective anisotropic strain shielding of a substrate-attached elongate cell (elongate, nominal 5 μm by 30 μm). As a result of its shape, a 2.5 μm deformation would produce 4 fold more strain when applied to the short versus the long axis of such a hypothetical cell. This assumes a nonaligned (isotropic) substrate. Clearly this would be enhanced if the matrix were also anisotropic.

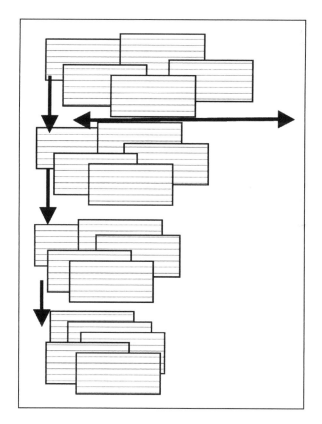

Figure 11. A general model or hypothesis for collagen fibrebundle or sheet 'slip' is illustrated. In this, overlapping collagen units carry uniaxial load (plane shown by the double headed arrow) due to bonding (nature unspecified) between the unit sheets or bundles. Geometric shortening of this system, in small increments, would result where units were able to slip between each other and rebond progressively in one direction (i.e., shortening by ratchet).[5,71] It is then possible to see how specialised contractile cells (e.g., (proto)-myofibroblasts) would provide that incremental, matrix-shortening direction.

mechanics, at the cell level, is rather special and a number of special factors operate. Perhaps the most important of these is the way in which the extra cellular matrix dictates the type and magnitude of cell deformation. This stress (or strain) shielding takes many forms but is certain to modify the predominant direction, frequency, and rate of change and overall magnitudes of cell loading. Musculo-skeletal cells also have a number of behavioural responses, affecting, (matrix) protein synthesis, shape/polarity and motility. Some of these are dictated directly by constraints or changes in mechanical properties of their surrounding ECM. The special feature of biological materials is that the resident cells are not passive, but themselves change the material properties, in a manner unique to material science.

In particular, anisotropic properties of collagen fibrous substrates can transmit/enhance important cytomechanical cues. However, by the same rule, inappropriate or disorganised fibre structures (e.g., scar tissue) would be expected to obscure and so degrade that same system of spatial cues, and with it the remodelling process.

Cytomechanics comprises a complex interplay of local (cell-level) forces and the material through which they operate. It is becoming increasingly clear that these controls can no longer be ignored nor treated inappropriately if we are to understand connective tissue control and dysfunction. The good news is that the rules/consequences of solid material mechanics (at any scale) are both well understood and mathematically predictable, in a way, which is not true for classical biological control systems.

Acknowledgements

I am grateful to Dr. Jaysheela Mudera for her pictures of scar tissues and the European Commission (Framework 5) BITES programme and BBSRC TIBS programme (31/E18398) for financial support.

References

1. Chen S, Springer TA. Selectin receptor-ligand bonds: Formation limited by shear rate and dissociation governed by the Bell model. Proc Natl Acad Sci USA 2001; 98:950-955.
2. Fulford GR, Katz DF, Powell RL. Swimming of spermatozoa in a linear viscoelastic fluid. Bioreheology 1998; 35:295-309.
3. Zippel H. Julius wolff and the law of bone remodelling. In: Regling G, ed. 'Wolff's Law and Connective Tissue Regulation'. Berlin: Walter de Gruyter, 1992:1-21.
4. Brown RA. Next generation tissue engineering: Clinical applications and mechanical control. In: Polak JM, Hench LL, Kemp P, eds. Future Strategies for Tissue and Organ Replacement. Singapore: World Scientific Publishing, 2002:48-75.
5. Tomasek JJ, Gabbiani G, Hinz B et al. Myofibroblasts and mechano-regulation of connective tissue remodelling. Nat Rev Mol Cell Biol 2002; 3:349-63.
6. Hukin DWL. Biomechanical properties of collagen. In: Weiss JB, Jayson MIV, eds. Collagen in Health and Disease. Edinburgh: Churchill-Livingston, 1982:49-72.
7. Frank C, Woo S, Andriacchi T et al. Normal ligament: Structure, function and composition. In: Woo S, Buckwalter JA, eds. Injury and Repair of Musculoskeletal Soft Tissues. Park Ridge: Am Acad Orthopaedic Surgeons, 1988:45-101.
8. Arnoczky S, Adams M, DeHaven K el al. Meniscus. In: Woo S, Buckwalter JA, eds. Injury and Repair of Musculoskeletal Soft Tissues. Park Ridge: Am Acad Orthopaedic Surgeons, 1988:487-537.
9. Eastwood M, Mudera VC, McGrouther DA et al. Effect of precise mechanical loading on fibroblast populated collagen lattices. Cell Motil Cytoskel 1998; 40:13-21.
10. Mudera VC, Pleass R, Eastwood M et al. Molecular responses of human dermal fibroblasts to dual cues: Contact guidance and mechanical load. Cell Motil Cytoskel 2000; 45:1-9.
11. Jackson DS. The dermal scar. In: Jayson MIV, Weiss JB, eds. Collagen in Health and Disease. Edinburgh: Churchill Livingston, 1982:466-474.
12. Woo SL-Y, Inoue M, McCurk-Burleson E et al. Treatment of the medial collateral ligament injury: II structure and function of canine knees in response to differing treatment regimes. Am J Sports Med 1987; 15:22-29.
13. Brown RA. Bioartificial Implants: Design and tissue engineering. In: Elices M, ed. Structural Biological Materials. Oxford: Elsevier Science, 2000:105-160.
14. Rheinwald JG, Green H. Serial cultivation of strains of human epidermal keratinocytes: The formation of keratinising colonies from single cells. Cell 1975; 6:331-34.

15. Falanga V, Margolis D, Alvarez O et al. Rapid healing of venous ulcers and lack of clinical rejection with an allogeneic cultured human skin equivalent. Arch Dermatol 1998; 134:293-300.
16. Rudolf R, Vandeberg J, Ehrlich HP. Wound contraction and scar contracture. In: Lindblad WJ, Cohen IK, Diegelmann RF, eds. Wound Healing: Biochemical and Clinical Aspects. Philadelphia: WB Saunders Co., 1992:96-114.
17. Muller P, Dahners LE. A study of ligamentous growth. Clinc Orthop Rel Res 1988; 229:274-277.
18. Nemetschek T et al. Functional properties f parallel fibrilled connective tissue with special regard to viscoelasticity (author's translation). Virchow. Arch A Pathol Anat Histol 1980; 386:125-151.
19. Brown RA, Byers PD. Swelling of cartilage and expansion of the collagen network. Calcif Tiss Int 1989; 45:260-261.
20. Hinz B, Gabbiani G. Mechanisms of force generation and transmission by myofibroblasts. Curr Opin Biotechnol 2003; 14:538-546.
21. Wallace AL, Hollinshead RM, Frank CB. Creep behaviour of a rabbit model of ligament laxity after electro thermal shrinkage in vivo. Am J Sport Med 2002; 30:98-102.
22. Bogatkevich GS, Tourkina E, Abrams CS et al. Contractile activity and smooth muscle alpha-actin organisation in thrombin-induced human lung myofibroblasts. Am J Phys Lung Cell Mol Physiol 2003; 285:L334-343.
23. Sethi KK, Mudera V, Sutterlin R et al. Contraction-mediated pinocytosis of RGD-peptide by dermal fibroblast: Inhibition of matrix attachment blocks contraction and disrupts microfilament organisation. Cell Moitil Cytoskel 2002; 52:231-241.
24. Sethi KK, Yannas IV, Mudera V et al. Sequential utilisation of fibronectin, vitronectin and collagen during fibroblast-mediated collagen contraction. Wound Rep Regen 2002; 10:397-408.
25. Choquet D, Felsenfeld DP, Sheetz MP. Extra cellular matrix rigidity causes strengthening of integrin-cytoskeletal linkages. Cell 1997; 88:39-48.
26. Parjapati RT, Eastwood M, Brown RA et al. Duration and alignment of mechanical stress are critical to activation of fibroblasts. Wound Repair Regeneration 2000; 8:239-247.
27. Kessler D, Dethlefsen S, Haase I et al. Fibroblasts in mechanically stressed collagen lattices assume a synthetic phenotype. J Biol Chem 2001; 276:36575-85.
28. Brown RA, Sethi KK, Gwanmesia I et al Enhanced fibroblast contraction of 3D collagen lattices and integrin expression by TGF-β1 and –β3: Mechanoregulatory growth factors? Exp Cell Res 2002; 274:310-322.
29. Delvoye P, Wiliquet P, Leveque JL et al. Measurement of mechanical forces generated by skin fibroblasts embedded in a 3D collagen gel. J Invest Dermatol 1991; 97:898-902.
30. Kolodney MS, Wysolmerski RB. Isometric contraction by fibroblasts and endothelial cells in tissue culture: A quantitative study. J Cell Biol 1992; 117:73-82.
31. Eastwood M, McGrouther DA, Brown RA. A culture force monitor for measurement of contraction forces generated by dermal fibroblast cultures: Evidence for cell-matrix mechanical signalling. Biochem Biophys Acta 1994; 1201:186-192.
32. Brown RA, Prajapati R, McGrouther DA et al. Tensional homeostasis in dermal fibroblasts: Mechanical responses to mechanical loading in three-dimensional substrates. J Cell Physiol 1998; 175:323-332.
33. Huang D, Chang TR, Aggarwal A et al. Mechanisms and dynamics of mechanical strengthening in ligament-equivalent fibroblast-populated collagen matrices. Ann Biomed Enã¬1993; 21:289-305.
34. Watatsuki T, Kolodney MS, Zahalak Gi et al. Cell mechanics studied by a reconstituted model tissue. Biophys J 2000; 79:2353-2368.
35. Grinnell F, Ho CH. Transforming growth factor beta stimulates fibroblast-collagen contraction by different mechanisms in mechanically loaded and unloaded matrices. Exp Cell Res 2002; 273:248-255.
36. Harris AK, Stopak D, Wild P. Fibroblast traction as a mechanism for morphogenesis. Nature 1981; 290:249-251.
37. Elsdale T, Bard J. Collagen substrata for studies on cell behaviour. J Cell Biol 1972; 54:626-637.
38. Fringer J, Grinnell F. Fibroblast quiescence in floating and released collagen matrices: Contribution of the ERK signalling pathway and actin cytoskeletal organisation. J Biol Chem 2000; 276:31047-52.
39. Cheema U, Yang SH, Mudera V et al. 3D in vitro model of early skeletal muscle development. Cell Motil Cytoskel 2003; 54:226-236.
40. Cheema U, Brown RA, Mudera V et al. Mechanical signals and IGF-1 gene-splicing in vitro in relation to development of skeletal muscle. J Cell Physiol 2005; 202:67-75.
41. Campbell BH, Clark WW, Wang JH. A multi-station culture force monitor system to study cellular contractility. J Biomech 2003; 36:137-140.
42. Pouliot Y, Gravel M, Holland PC. Developmental regulation of M-cadherin in the terminal differentiation of skeletal myoblasts. Dev Dyn 1994; 200:305-12.

43. Talas G, Adams TST, Eastwood M et al. Phenytoin reduces the contraction of recessive dystrophic epidermolysis bullosa fibroblast populated collagen gels. Int J Biochem Cell Biol 1997; 29:261-70.
44. Eastwood M, Porter RA, Khan U et al. Quantitative analysis of collagen gel contractile forces generated by dermal fibroblasts and the relationship to cell morphology. J Cell Physiol 1996; 166:33-42.
45. Clark RA, Lin F, Greiling D et al. Fibroblast invasive migration into fibronectin/fibrin gels requires a previously uncharacterised Dermatan sulphate CD44 proteoglycan. J Invest Derm 2004; 122:266-277.
46. Serini G, Gabbiani G. Mechanisms of myofibroblast activity and phenotype modulation. Exp Cell Res 1999; 250:273-283.
47. Desmouliere A, Geinoz A, Gabbiani F et al. TGFβ1 induces α-smooth muscle actin expression in granulation tissue myofibroblasts and in quiescent and growing cultured fibroblasts. J Cell Biol 1993; 122:103-111.
48. Arora PD, Narani N, McCullock CA. The compliance of collagen gels regulates TGFβ induction of α-smooth muscle actin in fibroblasts. Am J Path 1999; 154:871-882.
49. Bischofs IB, Schwarz US. Cell organisation in soft media due to mechanosensing. Proc Natl Acad Sci USA 2003; 100:9274-79.
50. Grinnell F. Fibroblasts, myofibroblasts and wound contraction. J Cell Biol 1994; 124:401-404.
51. Porter RA, Brown RA, Eastwood M et al. Ultra structural changes during contraction of collagen lattices by ocular fibroblasts. Wound Repair Regen 1998; 6:157-166.
52. Hinz B, Gabbiani G, Chaponnier C. The NH2 terminal of alpha smooth muscle actin inhibits force generation by the myofibroblast in vitro ad in vivo. J Cell Biol 2002; 13:657-663.
53. Kemp TJ, Sadusky TJ, Simon M et al. Identification of a novel stretch-responsive skeletal muscle gene (Smpx). Genomics 2001; 72:260-271.
54. Grinnell F. Fibroblast-collagen-matrix contraction: Growth factor signalling and mechanical loading. Trends Cell Biol 2000; 10:362-365.
55. Bishop JE, Butt R, Dawes K et al. Mechanical load enhances the stimulatory effect of PDGF on pulmonary artery fibroblast procollagen synthesis. Chest 1998; 114:25S.
56. Urban JP. The chondrocyte, a cell under pressure. Br J Rheum 1994; 33:901-8.
57. Buschmann MD, Gluzband YA, Grodzinsky AJ et al. Mechanical compression modulates matrix biosynthesis in chondrocyte/agarose culture. J Cell Sci 1995; 108:1497-508.
58. Fray TR, Molloy JE, Armitage MP et al. Quantification of single human dermal fibroblast contraction. Tissue Eng 1998; 4:281-291.
59. Curtis A, Wilkinson C. Nanotechniques and approaches in biotechnology. Trends Biotech 2001; 19:97-101.
60. Clark P, Connolly P, Curtis AS et al. Cell guidance by ultrafine topography in vitro. J Cell Sci 1991; 99:73-7.
61. Dalby MJ, Riehle MO, Johnstone H et al. In vitro reaction of endothelial cells to polymer demixed nanotopography. Biomaterials 2002; 23:2945-2954.
62. Curtis A, Wilkinson C. New depths in cell behaviour: Reactions of cells to nanotopography. Biochem Soc Symp 1999; 65:15-26.
63. Dalby MJ, Riehle MO, Sutherland DS et al. Use of nanotopography to study mechanotransduction in fibroblasts—methods and perspectives. Eur J Cell Biol 2004; 83:159-169.
64. Wakatsuki T, Elson EL. Reciprocal interaction between cells and extra cellular matrix during remodelling of tissue constructs. Biophys Chem 2003; 100:593-605.
65. Parsons M, Kessler E, Laurent GJ et al. Mechanical load enhances procollagen processing in dermal fibroblasts by regulating levels of procollagen C-proteinase. Exp Cell Res 1999; 252:319-31.
66. Garvin J, Qi J, Maloney M et al. Novel system for engineering bioartificial tendons and application of mechanical load. Tissue Eng 2003; 9:967-979.
67. Laxminarayanan K, Weiss JA, Wessman MD et al. Design and application of a test system for viscoelastic characterisation of collagen gels. Tissue Eng 2004; 10:241-252.
68. Sawhney RK, Howard J. Slow local movements of collagen fibres by fibroblasts drive the global self-organisation of collagen gels. J Cell Biol 2002; 157:1083-1091.
69. Prajapati RT, Chavally-Mis B, Herbage D. Mechanical loading regulates protease production by fibroblasts in 3-dimensional collagen substrates. Wound Repair Regen 2000; 8:227-238.
70. Rotwein P, Pollock KM, Didier DK et al. Organisation and sequence of the human IGF-1 gene. Alternative RNA spicing produces two IGF-1 precursor peptides. J Biol Chem 1986; 261:4828-4832.
71. Glimcher M, Peabody HM. In: McFarlane R, McGrouther DA, Flint MH, eds. Dupuytrens Disease. Edinburgh: Churchill Livingstone, 1990:72-85.
72. Marenzana M, Wilson-Jone N, Mudera V et al. Spatial remodelling of 3D collagen networks regulated by cell contraction and low frequency cyclical mechanical loading. Tissue Eng 2003; 9:798-799.

CHAPTER 2

Scleroderma Lung Fibroblasts:
Contractility and Connective Tissue Growth Factor

Galina S. Bogatkevich,* Anna Ludwicka-Bradley, Paul J. Nietert and Richard M. Silver

Introduction

In the pathogenesis of pulmonary fibrosis in general and systemic sclerosis (SSc, scleroderma) in particular, lung fibroblasts undergo specific phenotypic modulation and develop cytoskeletal features similar to those of smooth muscle cells. These phenotypically altered, activated fibroblasts, or myofibroblasts, express a contractile isoform of actin (α-smooth muscle actin) and promote contractility of lung parenchyma. Constitutively activated SSc fibroblasts produce an over abundance of collagen, fibronectin, and other extracellular matrix (ECM) proteins. They also overexpress several profibrotic receptors, including receptors for the dominant fibrogenic cytokine TGF-β. Recently, we found that the main receptor for thrombin on lung fibroblasts, Protease-Activated Receptor (PAR)-1, is also abundantly expressed in scleroderma lung tissue and, furthermore, its expression is observed in association with myofibroblasts. Moreover, thrombin itself is capable of differentiating normal lung fibroblasts to a myofibroblast phenotype. Such differentiation occurs via protein kinase C and a RhoA-dependent pathway.

Additionally, thrombin induces several potent fibrogenic cytokines. CTGF, whose expression is constitutively upregulated in scleroderma fibroblasts and correlates well with the severity of lung fibrosis, is one such potent fibrogenic cytokine induced by thrombin. This study was undertaken to establish the link between the expression of CTGF and the contractile activity observed in scleroderma lung fibroblasts.

Scleroderma Lung Fibrosis and Myofibroblasts

SSc is an autoimmune rheumatic disease that affects about 250,000 Americans, primarily females who are 30 to 50 years of age at disease onset. Currently, the leading cause of death in scleroderma patients is pulmonary dysfunction as a result of progressive interstitial lung fibrosis.[1] The conceptual process of lung fibrosis involves the presence of tissue injury, the release of fibrogenic factors, and the induction of myofibroblasts culminating in enhanced ECM deposition.[2-4] Myofibroblasts appear to be the principal mesenchymal cells responsible for tissue remodeling, collagen deposition, and the restrictive nature of the lung parenchyma associated with pulmonary fibrosis.[5-8]

We have demonstrated that myofibroblasts are present in the bronchoalveolar lavage fluid (BALF) of SSc patients and that myofibroblasts cultured from SSc BALF express more

*Corresponding Author: Galina S. Bogatkevich—Medical University of South Carolina, 96 Jonathan Lucas Street, Suite 912, Charleston, South Carolina 29425, U.S.A.
Email: bogatkev@musc.edu

Tissue Repair, Contraction and the Myofibroblast,
edited by Christine Chaponnier, Alexis Desmoulière and Giulio Gabbiani.
©2006 Landes Bioscience and Springer Science+Business Media.

collagen I, III, and fibronectin than normal lung fibroblasts.[9] They also show a greater proliferative response upon exposure to TGF-β and platelet-derived growth factor (PDGF) when compared to normal lung fibroblasts.[9,10] Recently we reported increased numbers of myofibroblasts in lung tissue from SSc where lesions with high ECM accumulation are present.[11] Several groups of investigators have demonstrated a correlation between fibrosis and α-smooth muscle actin expressing myofibroblasts in a number of different tissues.[5,8,12,13] Myofibroblasts isolated from various fibrotic tissues, including lungs, are thought to be the primary source of collagen and other ECM proteins.[5,9,11,12] Studies in animals employing the bleomycin-induced model of pulmonary fibrosis have identified myofibroblasts to be the primary source of increased collagen expression and a major source of cytokines and chemokines as well.[14,15]

TGF-β, Thrombin and CTGF in SSc Lung Fibrosis

The presence of myofibroblasts has been extensively documented in active fibrotic lesions in many diseases, including SSc lung disease.[6,8,9] However, the precise sources of such myofibroblasts are still not well known. Relative contributions from circulating mesenchymal stem cells or from local trans-differentiation of epithelial cells to fibroblasts, demonstrated in other organ systems,[16,17] have not been observed in the lung. It seems as if lung fibroblasts may differentiate to a myofibroblast phenotype under the influence of local growth factors and cytokines.[18] One such growth factor is TGF-β1, the most potent profibrotic cytokine and a powerful stimulator of the production of extracellular matrix components, including collagen, fibronectin, fibrillin and proteoglycans.[19]

We recently reported that thrombin also mediates differentiation of lung fibroblasts to a myofibroblast phenotype, apparently at an even earlier stage than TGF-β1.[11,20,21] We demonstrated that the expression of the main receptor for thrombin in lung fibroblasts, PAR-1, is dramatically increased in lung tissue from scleroderma patients, mainly in lung tissue containing inflammatory and fibroproliferative foci.[11] PAR-1 expression decreases in the later stages of pulmonary fibrosis, suggesting that its role takes place early in the development of lung fibrosis. PAR-1, which is responsible for most cellular events induced by thrombin, colocalizes with myofibroblasts in scleroderma lung tissue, lending additional support for thrombin and PAR-1 in the process of lung fibroblast activation.[11]

Another protein essential for matrix synthesis and remodeling in scleroderma is CTGF. This growth factor is constitutively overexpressed by fibroblasts in fibrotic skin lesions as well as in lung fibroblasts from patients with systemic sclerosis.[22-24] The majority of the CTGF generated by scleroderma fibroblasts is associated with the cell layer, suggesting that CTGF acts in an autocrine fashion.[22,24] CTGF regulates deposition of ECM components in fibroblasts and is responsible for the persistent fibrotic reaction. Increased amounts of CTGF are found in scleroderma patients with more extensive skin involvement and a greater severity of fibrosis.[23] In addition to increased tissue expression, CTGF levels are increased in the serum of patients with scleroderma.[23] Higher levels of CTGF are seen in the diffuse form of scleroderma than in the limited form.[23] BALF from systemic sclerosis patients with active lung fibrosis contains much higher levels of CTGF compared to patients without fibrotic lung pathology.[23] We found that thrombin at physiological levels promotes overexpression of CTGF in normal lung fibroblasts approximately 8-fold, compared to only 3-fold induction by TGF-β (Fig. 1). Similar results were reported from Laurent's laboratory.[25] Lung fibroblasts derived from scleroderma patients inherently contain approximately 5.5-fold protein level of CTGF compared to normal lung fibroblasts (Fig. 1).

Since thrombin and TGF-β each induce CTGF production in fibroblasts, it has been suggested that CTGF may be a common downstream mediator for pathways associated with activation of both thrombin and TGF-β.[26]

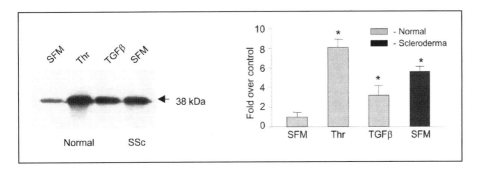

Figure 1. CTGF protein level in normal and SSc lung fibroblasts treated with thrombin or TGF-β. Confluent lung fibroblasts were cultured in serum-free conditions for 24 hours followed by incubation for 24 additional hours in serum-free medium (SFM), thrombin (1 U/ml), or TGF-β$_1$ (5 ng/ml). The cells were harvested in PBS and solubilized in heparin-sepharose binding buffer containing 100 mM Tris, pH 7.4; 10 mM trisodium citrate; 150 mM NaCl; 1% Triton. CTGF protein was selected with heparin Sepharose (Amersham Biosciences), subjected to 10% polyacrylamide gel, and analyzed by Western blot with anti-CTGF antibody (R and D Systems, Inc). CTGF bands were quantified by densitometry using NIH Image software. The experiment was performed three times and mean values ± SD is presented. The asterisk represents statistically significant differences (p < 0.05) between normal lung fibroblasts stimulated with thrombin or TGF-β versus nonstimulated cells and between SSc lung fibroblasts versus normal lung fibroblasts.

Contractile Activity of CTGF

Contractility, a predominant feature of myofibroblasts, is equally promoted by either thrombin or TGF-β.[19-21] Contractile forces of the myofibroblast are generated by α-smooth muscle actin, which is extensively expressed in stress fibers and by large fibronexus adhesion complexes connecting intracellular actin with extracellular fibronectin fibrils.[27] Fibroblasts cultured in collagen matrices have been used to develop in vitro models of fibrocontractile diseases.[28] Studies in our laboratory have focused on stressed matrix contractions as an equivalent of myofibroblast contraction in vivo.[20,21] We observed that SSc lung fibroblasts contain an abundant amount of α-smooth muscle actin and, accordingly, are capable of contracting collagen gels without exogenous treatment.[20] However, distinct SSc cell lines contract collagen gels to different degrees (Fig. 2A). To investigate the correlation between contractile activity and expression of CTGF by SSc lung fibroblasts, we measured the amount of CTGF using anti-CTGF antibody in collagen gels undergoing contraction by scleroderma lung fibroblasts. Collagen gels were digested with collagenase, and lung fibroblasts released from collagen were resolved on SDS-PAGE. CTGF bands were visualized by Western blot with anti-CTGF antibody and quantified by densitometry using NIH Image software. The association of the CTGF expression with ability of SSc lung fibroblasts to contract collagen gel was tested using Spearman rank correlation test. We found that cell lines containing more CTGF demonstrated greater contractile activity when compared to the cells containing less CTGF. For example, SSc cell line 3 having the highest CTGF level (895 ± 62 densitometric value) contracted collagen gels from ~15 mm in diameter to less then 6 mm in diameter (~9 mm contraction). In contrast, another SSc cell line expressing lower CTGF levels (412 ± 97 densitometric value) produced contraction of only about 5mm (Fig. 2).

To investigate if CTGF can induce contractile activity by lung fibroblasts, we transfected normal lung fibroblasts with human CTGF in V5/His/pcDNA3.1 and then performed collagen gel contraction assays. We observed that lung fibroblasts transfected with CTGF contracted gels from 15 mm to ~7 mm diameter (~8 mm contraction), whereas nontransfected cells or cells transfected with vector as a control did not contract collagen gels (Fig. 3). To further confirm that CTGF can induce contractile activity by lung fibroblasts, we performed

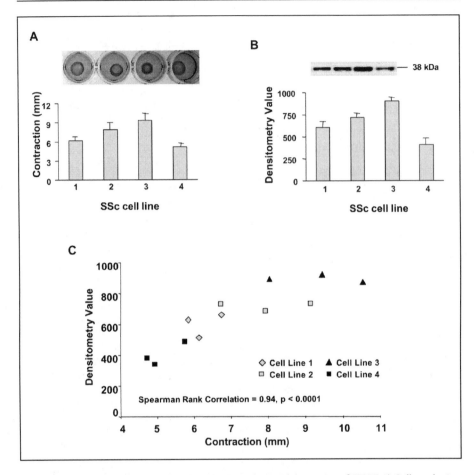

Figure 2. Contractile activity of SSc lung fibroblasts correlates with expression of CTGF. A) Collagen lattices were prepared using type I collagen from rat tail tendon as previously described.[20,21] SSc lung fibroblasts were suspended in collagen (2.5 x 10[5] cells/ml final concentration) and aliquoted into 24-well plates. Collagen lattices were polymerized for 45 min at 37°C followed by incubation with medium containing 10% fetal calf serum for 4 h, followed by overnight incubation in serum-free medium. To initiate collagen gel contraction, polymerized gels were gently released from underlying culture dish. The diameters of the gels were measured immediately and 24 h after release of gels. The degree of contraction was determined as difference between initial diameter of the gel and diameter of the gel in 24 h. B) Collagen gels were collected, digested with collagenase, and analyzed by Western blot using anti-CTGF antibody. CTGF bands were quantified by densitometry using NIH Image software. The experiment was performed three times and mean values ± SD is presented. C) The correlation between the level of CTGF and ability of distinct SSc cell lines to contract collagen gel was analyzed by using the Spearman rank correlation.

collagen gel contraction assays utilizing normal lung fibroblasts treated with recombinant CTGF. Human CTGF in V5/His/pcDNA3.1 vector was purified from transfected He-La cells using ProBond nickel-chelating resin and heparin sepharose. We observed that CTGF induced collagen gel contraction in a dose-dependent manner with maximum effects at 100 ng/ml and 1 µg/ml (Fig. 4A). Collagen gels contracted from 15 mm in diameter to ~9 mm (~6 mm contraction) within 24 hours of CTGF stimulation.

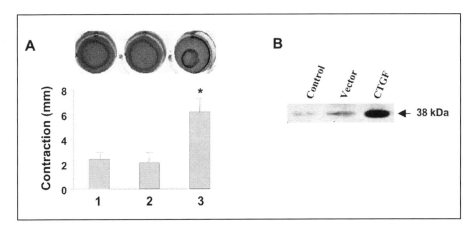

Figure 3. Overexpression of CTGF in normal lung fibroblasts induces collagen gel contraction. A) Normal lung fibroblasts were transiently transfected with CTGF DNA and cultured in collagen gel on 24-well plates. Polymerized gels were released from the underlying culture dish 24 h after transfection and incubated in serum-free medium for another 24 h. Column 1 represents nontransfected cells as a control, 2—cells transfected with vector, 3—cells transfected with CTGF. The experiment was performed three times and mean values ± SD is presented. The asterisk represents statistically significant differences (p < 0.05) between cells transfected with CTGF versus nontransfected cells. B) Collagen gels were collected, digested with collagenase, and analyzed by Western blot using anti-CTGF antibody.

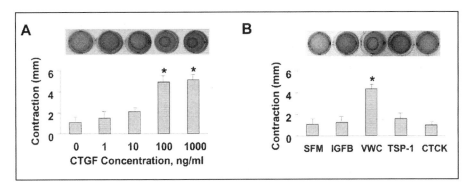

Figure 4. Recombinant CTGF induces collagen gel contraction by lung fibroblasts via VWC domain. Normal lung fibroblasts were cultured in collagen lattices as described earlier. A) CTGF (concentrations from 1 ng/ml to 1 μg/ml) was added and collagen gel contraction was measured after 24 h of incubation. B) Normal lung fibroblasts were left in serum-free medium (SFM) or stimulated with 100 ng/ml of IGFB, VWC, TSP-I, and CTCK domains of CTGF for 24 h. Collagen gel contraction was measured as stated above. Data represent mean values ± SD of three experiments, each performed in duplicate. The asterisk corresponds to statistically significant differences (p < 0.05) between cells stimulated with whole CTGF (A) or CTGF domain (B) versus nonstimulated cells.

Next, we initiated a series of experiments to identify the structural module of CTGF responsible for contractile activity. Protein structural modules (also known as modular protein) are well demarcated and independently folded portions of proteins. Such domains are noncatalytic and bind specifically to short continuous peptide sequences in their binding partner(s) via one or more ligand-binding surfaces.[29] Modular protein domains are autonomous in the sense that in most cases they may be removed from the original parent protein without compromising their ability to bind their cognate or target peptide ligands. Human CTGF protein contains 349 amino acids organized into four distinct structural modules following the signal peptide: insulin-like growth-factor-binding domain (IGFB) contains 63 amino acids (Gly31-Gly93); von Willebrand factor type C (VWC) domain or chordin-like domain[30] contains 60 amino acids (Cys103-Glu162); thrombospondin type I repeat (TSP1) contains 41 amino acids (Gln202-Cys242); and carboxyl-terminal cysteine knot-like domain (CTCK) contains 70 amino acids (Lys261-Pro330). The CTCK domain of CTGF is present in biological fluid and is sufficient for some biological activity.[31] Recently, Ball et al reported the functional activity of the CTCK domain produced as a maltose binding fusion protein in *E. coli*. After removal of the fusion part, recombinant CTCK promoted dose-dependent adhesion of fibroblasts, myofibroblasts, endothelial cells and epithelial cells.[32]

To determine whether any CTGF domains can autonomously induce contractile activity, they were cloned as glutathione S-transferase (GST)-fused proteins into pGEX2T vector and purified using Glutathione Sepharose. For collagen gel contraction assay by lung fibroblasts we used IGFB, VWC, TSP1, and CTCK domains of CTGF. We found that the VWC domain of CTGF in a concentration of 100 ng/ml demonstrated profound contractile activity, whereas IGFB, TSP1 and CTCK had no effect on collagen gel contraction (Fig. 4B). Collagen gels contracted from -15 mm in diameter to -10 mm (-5 mm contraction) within 24 hours after VWC stimulation, but remained unchanged after stimulation with IGFB, TSP1, or CTCK domains of CTGF.

Conclusions

The role of myofibroblasts in various fibrotic disorders is currently well established. These smooth-muscle-like fibroblasts promote deposition of ECM proteins and contractility of lung parenchyma. The present studies were performed to characterize the contractile activity of SSc lung fibroblasts. Previously, we demonstrated that the early stages of interstitial lung disease of SSc are characterized by a prominence of cells possessing a myofibroblast phenotype. A major feature of such myofibroblasts is contractility, explained by an over-expression of α-smooth muscle actin. Here, we demonstrate for the first time that the contractility of SSc lung fibroblasts depends on expression of CTGF as well, and that the VWC domain is primarily responsible for the contractile activity of CTGF in human lung fibroblasts. Future studies are required to identify the mechanisms by which CTGF stimulates collagen gel contraction.

Acknowledgements

National Institutes of Health Grant P60 AR049459-01 (Multidisciplinary Clinical Research Center, to R.M. Silver) and NRSA 1 F32 HL 69689-02 (to G.S. Bogatkevich), Grants from the Scleroderma Foundation (to A. Ludwicka-Bradley), and Pilot funding from the MUSC Office of the Provost (to G.S. Bogatkevich) supported this work.

References

1. Silver RM, Bolster MB. Systemic sclerosis (scleroderma). In: Austen KF, Frank MM, Atkinson JP et al eds. Samter's Immunologic Diseases. 6th ed. Philadelphia: Lippincott Williams and Wilkins, 2001:504-519.
2. Steen VD, Medsger Jr TA. Severe organ involvement in systemic sclerosis with diffuse scleroderma. Arthritis Rheum 2000; 43:2437-2444.
3. Ward PA, Hunninghake GW. Lung inflammation and fibrosis. Am J Respir Crit Care Med 1998; 157:S123-S129.
4. Bouros D, Wells AU, Nicholson AG et al. Histopathologic subset of fibrosing alveolitis in patients with systemic sclerosis and their relationship to outcome. Am J Respir Crit Care Med 2002; 165:1581-6.
5. Tomasek JJ, Gabbiani G, Hinz B et al. Myofibroblasts and mechano-regulation of connective tissue remodeling. Nature Rev Mol Cell Biol 2002; 3:349-363.

6. Pache JC, Chrstakos PG, Gannon DE et al. Myofibroblasts in diffuse alveolar damage of the lung. Modern Pathol 1998; 11:1064-70.
7. Low RB. Modulation of myofibroblast and smooth-muscle phenotypes in the lung. Curr Top Pathol 1999; 93:19-26.
8. Kapanci Y, Gabbiani G. Contractile cells in pulmonary alveolar tissue. In: Crystal RG, West JB, eds. The lung: Scientific foundation. Philadelphia: Lippincott-Raven, 1997:697-707.
9. Ludwicka A, Trojanowska M, Smith EA et al. Growth and characterization of fibroblasts obtained from bronchoalveolar lavage of scleroderma patients. J Rheumatol 1992; 19:1716-1723.
10. Ludwicka A, Ohba T, Trojanowska M et al. Elevated levels of TGF-β1 and PDGF in scleroderma bronchoalveolar lavage fluid. J Rheumatol 1995; 22:1876-1883.
11. Ludwicka-Bradley A, Bogatkevich GS, Silver RM. Thrombin-mediated cellular events in pulmonary fibrosis associated with systemic sclerosis (scleroderma). Clin Exp Rheumatol 2004; 22:S38-S46.
12. Zhang K, Rekhter MD, Gordon D et al. Myofibroblasts and their role in lung collagen gene expression during pulmonary fibrosis: A combined immunohistochemical and in situ hybridization. Am J Pathol 1994; 145:114-25.
13. Walker GA, Guerrero IA, Leinwand LA. Myofibroblasts: Molecular crossdressers. Curr Top Dev Biol 2001; 51:91-107.
14. Zhang H, Gharaee-Kermani M, Zhang K et al. Lung fibroblast α-smooth muscle actin expression and contractile phenotype in bleomycin-induced pulmonary fibrosis. Am J Pathol 1996; 148:527-37.
15. Vyalov SL, Gabbiani G, Kapanci Y. Rat alveolar myofibroblasts acquire α-smooth muscle actin expression during bleomycin-induced pulmonary fibrosis. Am J Pathol 1993; 143:1754-1765.
16. Pittenger MF, Mackay AM, Beck SC et al. Multilineage potential of adult human mesenchymal stem cells. Science 1999; 284:143-147.
17. Iwano M, Plieth D, Danoff TM et al. Evidence that fibroblasts derive from epithelium during tissue fibrosis. I Clin Invest 2002; 110:341-350.
18. Phan SH. The myofibroblast in pulmonary fibrosis. Chest 2002; 122:286S-289S.
19. Vaughan MB, Howard EW, Tomasek JJ. Transforming growth factor-β1 promotes the morphological and functional differentiation of the myofibroblast. Exp Cell Res 2000; 257:180-189.
20. Bogatkevich GS, Tourkina E, Silver RM et al. Thrombin differentiates normal lung fibroblasts to a myofibroblast phenotype via the proteolytically activated receptor-1 and a protein kinase C-dependent pathway. J Biol Chem 2001; 276:45184-92.
21. Bogatkevich GS, Tourkina E, Abrams CS et al. Contractile activity and smooth muscle-α actin organization in thrombin-induced human lung fibroblasts. Am J Physiol Lung Cell Mol Physiol 2003; 285:L334-L343.
22. Leask A, Sa S, Holmes A et al. The control of ccn2 (CTGF) gene expression in normal and scleroderma fibroblasts. J Clin Pathol Mol Pathol 2001; 54:180-183.
23. Sato S, Nagaoka T, Hasegawa M et al. Serum levels of connective tissue growth factor are elevated in patients with systemic sclerosis: Association with extent of skin sclerosis and severity of pulmonary fibrosis. J Rheumatol 2000; 27:149-154.
24. Shi-wen X, Pennington D, Holmes A et al. Autocrine overexpression of CTGF maintains fibrosis: RDA analysis of fibrosis genes in systemic sclerosis. Exp Cell Res 2000; 259:213-224.
25. Chambers RC, Leoni P, Blanc-Brude OP et al. Thrombin is a potent inducer of connective tissue growth factor production via proteolytic activation of protease-activated receptor-1. J Biol Chem 2000; 275:35584-35591.
26. Leask A, Holmes A, Abraham DJ. Connective tissue growth factor: A new and important player in the pathogenesis of fibrosis. Current Rheumatol Reports 2002; 4:136-142.
27. Gabbiani G. The myofibroblast in wound healing and fibrocontractive diseases. J Pathol 2003; 200:500-503.
28. Grinnell F. Signal transduction pathways activated during fibroblast contraction of collagen matrices. Curr Top Pathol 1999; 93:61-73.
29. Pawson T, Nash P. Assembly of cell regulatory systems through protein interaction domains. Science 2003; 300:445-452.
30. Abreu JG, Keptura NI, Reverside B et al. Connective tissue growth factor (CTGF) modulates cell signalling by BMP and TGFβ. Nature Cell Biol 2002; 4:599-604.
31. Brigstock DR, Steffen CL, Kim GY et al. Purification and characterization of novel heparin-binding growth factors in uterine secretory fluids. Identification as heparin-regulated M_r 10,000 forms of connective tissue growth factor. J Biol Chem 1997; 272:20275-20282.
32. Ball DK, Rachfal AW, Kemper SA et al. The heparin-binding 10 kDa fragment of connective tissue growth factor (CTGF) containing module 4 alone stimulates cell adhesion. J Endocrinol 2003; 176:R1-R7.

CHAPTER 3

Functional Assessment of Fibroblast Heterogeneity by the Cell-Surface Glycoprotein Thy-1

Carolyn J. Baglole, Terry J. Smith, David Foster, Patricia J. Sime, Steve Feldon and Richard P. Phipps*

Abstract

Fibroblasts are a heterogeneous population of structural cells whose primary function is the production of extracellular matrix for normal tissue maintenance and repair. However, fibroblasts provide much more than structural support as they synthesize and respond to many different cytokines and lipid mediators and are intimately involved in the processes of inflammation. It is now appreciated that fibroblasts exhibit phenotypic heterogeneity, differing not only between organ systems, but also within a given anatomical site. Subtypes of fibroblasts can be identified by the expression of markers such as Thy-1, a cell surface glycoprotein of unknown function. Initial characterization of fibroblasts as Thy-1+ or Thy-1− can be performed by immunofluorescence or flow cytometry. They can be sorted according to their expression of Thy-1 by fluorescence-activated cell sorting (FACS), cloning and/or magnetic beading, yielding greater than 99% purity. Fibroblasts that are separated into Thy-1+ and Thy-1− subsets exhibit differences in their morphological, immunological and proliferative responses and ability to differentiate into α-smooth muscle actin-expressing myofibroblasts and adipocyte-like lipofibroblasts, key cells for wound healing and fibrotic disorders. The identification of Thy-1 as a surface marker by which to separate fibroblast subtypes has yielded vital insight into diseases such as scarring and wound healing and highlights the concept of fibroblast heterogeneity. Future research into fibroblast subsets may lead to the tissue-specific treatment of disease such as idiopathic pulmonary fibrosis and Graves' ophthalmopathy.

Introduction

Fibroblasts are a heterogeneous population of structural cells whose primary function is the production of extracellular matrix (ECM) for tissue maintenance and repair.[1] Thus, fibroblasts play a pivotal role not only in maintaining tissue integrity, but also in healing processes. They participate in fibrotic (scarring) disorders[2] in lung, skin and other tissues.[3] Wound healing is a reparative process that results in the restoration of tissue with minimal loss of function.[3] Conversely, fibrosis is a response to tissue injury and the ensuing inflammatory response, which

*Corresponding Author: Richard P. Phipps—University of Rochester School of Medicine and Dentistry, Department of Environmental Medicine, 601 Elmwood Ave, Box 850, Rochester, New York 14642, U.S.A. Email: Richard_Phipps@urmc.rochester.edu

Tissue Repair, Contraction and the Myofibroblast,
edited by Christine Chaponnier, Alexis Desmoulière and Giulio Gabbiani.
©2006 Landes Bioscience and Springer Science+Business Media.

ultimately results in abnormal ECM production through the activation of fibroblasts.[1] Fibroblasts participate in fibrosis by differentiating into cells called myofibroblasts,[4,5] production of ECM and recruitment of lymphocytes to the site of inflammation.[6,7] These effector functions are mediated through the production of proinflammatory cytokines and prostaglandins. It is now well-recognized that fibroblasts exhibit phenotypic heterogeneity, differing not only between organ systems but also within an anatomical site. Subtypes of fibroblasts can be identified by morphological and functional characteristics, proliferative potential, biosynthetic capacity and the expression of surface markers such as C1q receptors[8] and Thy-1.[2,9-12]

Thy-1 is a 25-kD cell surface glycoprotein belonging to the immunoglobulin-like superfamily. Originally described as a marker of thymocyte differentiation in mice[13] it is expressed on subsets of neurons, epidermal cells, lymphocytes,[9,14,15] endothelial cells[13] and has been found on fibroblasts of all species studied thus far.[16,17] Fibroblasts can be derived from tissues such as lung, skin, orbit of the eye, spleen and cornea and can be separated on the basis of Thy-1 surface expression. Following establishment of a primary culture, the fibroblastic phenotype is documented using morphological characteristics and immunohistochemistry. Further characterization of the cells as expressing (Thy-1$^+$) or lacking Thy-1 (Thy-1$^-$) can be performed by immunofluorescence and flow cytometry. They may be sorted according to their expression of Thy-1 by fluorescence-activated cell sorting (FACS) and/or magnetic beading, yielding greater than 99% purity. Once separated, pure Thy-1$^+$ or Thy-1$^-$ fibroblast subsets can be propagated. Figure 1, provides an overview of fibroblast derivation and characterization (the reader is referred to ref. 18 for a detailed protocol).

Initial studies characterized fibroblast subsets obtained from the mouse lung[9,19] and demonstrated that they possessed identifying morphological characteristics. Thy-1$^+$ fibroblasts are generally spindle-shaped and synthesize large amounts of type I and III collagen. In contrast, Thy-1$^-$ fibroblasts are rounded, more spread and synthesize less collagen.[19] Induction of collagen synthesis by IL-4 is also greater in Thy-1$^+$ fibroblasts.[20] Fibroblasts from several species and varying anatomic regions have been distinguished with respect to the expression of Thy-1.[9,11,12,19,21-25] Although its precise function remains enigmatic, it is becoming increasingly clear that the level of Thy-1 expression correlates with distinct morphological and functional characteristics. These may play a role in the initiation and progression of inflammatory diseases such as pulmonary fibrosis, reproductive tract scarring and thyroid-associated ophthalmopathy.

Immunological and Inflammatory Characteristics of Thy-1 Fibroblast Subsets

The activation of fibroblasts by inflammatory stimuli results in their migration, proliferation and deposition of extracellular matrix components, important features involved in both wound healing and fibrosis. The differential expression of Thy-1 on fibroblast subsets may have important immunological consequences due, in part, to the inducible expression of MHC class II molecules, whose primary function is to present antigen to T cells to elicit an immune response. Phipps et al[9] initially demonstrated that mouse lung Thy-1$^-$, but not Thy-1$^+$ fibroblasts, upregulated class II MHC molecules when treated with interferon-γ (IFN-γ). Further, coculture of IFN-γ-treated Thy-1$^-$ fibroblasts (but not untreated Thy-1$^-$ fibroblasts) with T helper cells (in the presence of antigen) results in the proliferation of these T cells. Therefore, the Thy-1$^-$ fibroblast subset can present antigen to T cells, strongly suggesting that Thy-1$^-$ fibroblasts are involved in chronic lung inflammation and the development of lung fibrosis.[9] The ability of IFN-γ to upregulate MHC II molecules only on Thy-1$^-$ fibroblasts is not limited to mouse lung fibroblasts. Similar observations have been made in mouse splenic and human orbital fibroblasts where, only Thy-1$^-$ fibroblasts are able to upregulate class II MHC.[22,26]

In addition to the upregulation of MHC class II molecules on specific fibroblast subsets, the differential production of cytokines by Thy-1$^-$ versus Thy-1$^+$ fibroblasts has been examined. IL-6 is a cytokine upregulated during inflammation that has been implicated in the pathogenesis of inflammatory conditions such as pulmonary fibrosis[27] and Graves' disease.

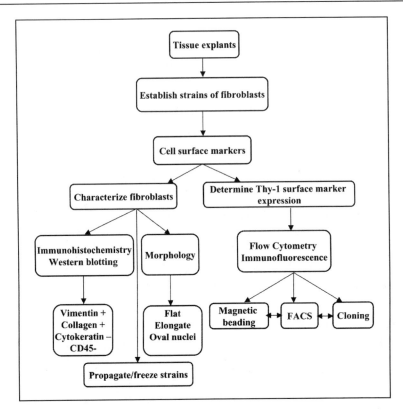

Figure 1. Strategies to establish and characterize fibroblast subsets. Tissue explants used to derive fibroblast strains from organs such as the lung are characterized by cell surface markers and morphological parameters. Fibroblasts express vimentin and collagens, but not cytokeratin (epithelial cell marker) and CD45 (blood cell marker). Thy-1 surface expression of the parental fibroblast strain can be determined by flow cytometry or immunofluorescence. Thy-1$^+$ and Thy-1$^-$ subsets can be sorted using fluorescence-activated cell sorting (FACS), magnetic beading and/or limit dilution cloning. Detailed protocols can be found in reference 18.

The latter is an autoimmune disease associated with localized manifestations, such as ophthalmopathy.[28] Both Thy-1$^+$ and Thy-1$^-$ fibroblasts subsets from the human orbit produced IL-6.[29] However, Thy-1$^-$ fibroblasts isolated from the rat lung exhibited increased levels of IL-6 mRNA in response to IL-1β.[24] This differential induction in IL-6 may reflect tissue and/or species-specific fibroblast heterogeneity.

In contrast to the uniform production of IL-6, IL-1β-activated Thy-1$^-$ fibroblasts from orbital tissue produce significantly more IL-8 than Thy-1$^+$ cells. IL-8 can activate T cells and neutrophils and is found at sites of tissue injury and chronic inflammation.[26,30] Collectively, these studies suggest that Thy-1$^-$ fibroblasts can act as antigen presenting cells and participate in the initial recruitment of inflammatory cells such as T cells and monocytes into tissue sites important in the initiation of wound healing and progression to fibrosis.

Another important regulator of inflammation is prostaglandin E$_2$, (PGE$_2$) one of the principal mediators of inflammatory conditions such as interstitial lung disease.[31] PGE$_2$, highly produced by fibroblasts, participates in leukocyte recruitment and synergizes with IL-8 for neutrophil recruitment. PGE$_2$ is produced through the action of cyclooxygenase (COX) enzymes, of which there are two isoforms, COX-1 and COX-2. In most tissues, COX-1 is constitutively expressed and participates in tissue homeostasis. Conversely, COX-2 is an immediate-early gene that usually is expressed at very low levels, but is rapidly induced following proinflammatory stimulation with cytokines such as with IL-1β.[12,32] In both human orbital[26] and myometrial fibroblasts,[12] Thy-1+ fibroblasts synthesize more PGE$_2$ following IL-1β treatment than do Thy-1- fibroblasts. Similar results were obtained with fibroblast subsets isolated from the female reproductive tract. Koumas et al[12] showed that PGE$_2$ production in the Thy-1- subset was largely unaffected by IL-1β treatment (and therefore very low) whereas production of PGE$_2$ increased two-fold in Thy-1+ cells. It was also found that COX-1 was highly expressed in untreated Thy-1+ fibroblasts and IL-1β upregulated COX-2.

Thy-1+, but not Thy-1-, fibroblasts are also able to upregulate the expression of CD40 following IFN-γ treatment.[11] CD40 was originally described as a receptor responsible for the activation and differentiation of B-lymphocytes.[3,33,34] CD40 is an ~50 kDa plasma membrane-spanning receptor and member of the tumour necrosis family (TNF)-α receptor superfamily. It is engaged by its ligand CD154 to promote cell survival, costimulatory protein (e.g., B7-1, B7-2) expression necessary for interaction with T-lymphocytes and B-cell immunoglobulin class switching. CD154 (also called CD40 ligand) up-regulates the synthesis of proinflammatory cytokines (IL-1, IL-6) and chemokines such as IL-8 and monocyte chemoattractant protein-1 (MCP-1) and PGE$_2$.[31,35] Thy-1+ myometrial fibroblasts treated with IFNγ and CD154 exhibit significantly increased IL-6, IL-8 and MCP-1 production.[11] Interestingly, the expression of CD40 occurred in a bimodal pattern in fibroblast subsets isolated from the human orbit.[26] IFN-γ treatment of isolated Thy-1+ and Thy-1- orbital fibroblasts increased CD40 expression in a portion of each subset. Functionally, CD40 ligation led to PGE$_2$ production in both subsets, although Thy-1+ cells were able to produce significantly more PGE$_2$. This observation suggests that the Thy-1+ subset plays an important role in scarring disorders in the orbit of the eye and lung through increased production of PGE$_2$.

Although T cells provide an important source of CD154, it was recently demonstrated that human lung fibroblasts express intracellular CD154 and that fibroblasts strains from patients with idiopathic pulmonary fibrosis express higher levels of CD154 than fibroblasts isolated from normal tissue.[3] Moreover, the expression of CD154 was heterogeneous in that there was a fraction of fibroblasts that did not stain positively.[3] Although the level of Thy-1 expression was not examined in parallel with CD154 expression in this study, it is possible that the heterogeneous nature of CD154 expression is related to the presence, or absence, of Thy-1. Thus, fibroblasts play an important role in the propagation of inflammatory processes through the autocrine/paracrine actions of the CD40-CD154 system.[3]

Fibrogenic and Proliferative Characteristics of Thy-1+ and Thy-1- Subsets

Fibroblasts, given the appropriate microenvironment, can acquire some of the characteristics of smooth muscle cells, defined by the expression of α-smooth muscle actin (α-SMA).[25] These differentiated fibroblasts, termed "myofibroblasts", have features that are intermediate to both fibroblasts and smooth muscle cells.[1] Although they occur in normal tissue, myofibroblasts are critically important in wound healing and fibrotic disorders[4,25,36,37] and are active producers of cytokines and collagen.[15] Alterations in fibroblast proliferation and increased density of myofibroblasts have important consequences in idiopathic pulmonary fibrosis, a poorly-treatable fibrotic lung disorder with a five-year mortality of more than 50%.[38-41] The ability of fibroblasts to acquire the myofibroblastic phenotype and thus participate in the progression of fibrosis correlates well with Thy-1 expression. We have shown that only human myometrial (Fig. 2) and orbital fibroblast subsets expressing Thy-1+ are capable of differentiating to myofibroblasts when treated with transforming growth factor-β (TGF-β),[25] a key mediator in fibrosis and derivation of the myofibroblastic phenotype. Interestingly, human myometrial Thy-1+ fibroblasts express low levels of α-SMA constitutively, unlike Thy-1+ fibroblasts from the orbit. This difference in basal α-SMA expression further reflects the tissue-specific

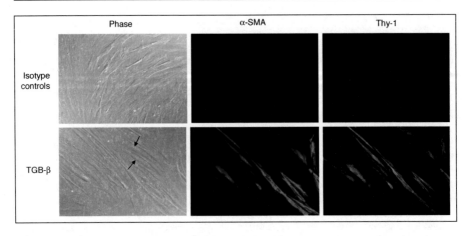

Figure 2. α-SMA is expressed in Thy-1⁺ human myometrial fibroblasts. Primary myometrial fibroblasts were double stained by immunofluorescence for α-SMA (PE; red) and Thy-1 (FITC; green) after a 4-day TGF-β treatment (5 ng/ml). Isotype controls for α-SMA and Thy-1 were mIgG2a and mIgG1 respectively, and stained negative for PE or FITC. Fibroblasts negative for Thy-1 were also negative for α-SMA. They are indicated in the phase picture by arrows. Original magnification x20. Reprinted from Am J Pathol 2003; 163:1291-1300, with permission from the American Society for Investigative Pathology.

fibroblast heterogeneity and may indicate the requirement for myofibroblasts in maintaining the normal architecture and function of the human myometrium.

The results presented in the study by Koumas et al[25] are in contrast to a recent study where Zhou and colleagues[15] separated rat lung fibroblasts on the basis of Thy-1 expression. They treated subsets with profibrotic cytokines such as IL-1β, IL-4 and platelet-derived growth factor (PDGF). Under basal conditions, Thy-1⁻ fibroblasts expressed more α-SMA than did Thy-1⁺ cells and when stimulated with these cytokines, Thy-1⁻ cells expressed significantly increased α-SMA levels. It was concluded that, at least in the rat lung, the presence of Thy-1 on the fibroblast surface confers resistance to the myofibroblastic phenotype.[15] These findings fully support the concept of tissue and/or species-specific fibroblast heterogeneity where differences in function are linked to anatomical location.

PDGF is an important mitogen and chemoattractant for cells of mesenchymal origin.[42] Moreover, its ability to activate fibroblasts appears to be related to the level of Thy-1 expression. For example, PDGF induces a myofibroblast phenotype in Thy-1⁻ rat lung fibroblasts[15] through differences in PDGF α-receptors (PDGFR-α).[43] There are three isoforms of PDGF; PDGF-AA, PDGF-BB and PDGF-AB- and their biological activity is determined by the expression of PDGFR-α and β-receptors (PDGFR-β).[42] Thy-1⁻ cells isolated from the rat lung proliferate in response to PDGF-AA whereas Thy-1⁺ fibroblasts do not.[43] This response correlated with expression levels of PDGFR-α, the receptor that confers the signal of PDGF-AA. In a study by Hagood and colleagues,[43] PDGFR-α levels were higher in Thy-1⁻ cells whereas PDGFR-β levels were similar in the two subsets. Further, PDGF-AA increased *c-myc* mRNA, an indicator of intracellular signaling, in Thy-1⁻ fibroblasts.[43]

The response to PDGF-AA and PDGFR-α levels in the Thy-1⁻ fibroblasts may result from differential responses to IL-1β, a regulator of PDGFR-α[44] and a potent inducer of inflammation. Hagood and colleagues recently reported that IL-1β increases PDGFR-α levels and proliferation in Thy-1⁻ cells and the IL-1 receptor antagonist (IL-1 RA) inhibits these processes.[24] Interestingly, other signaling components such as IL-1β binding, activation of p38 MAP kinases and IL receptor subtypes were similar between Thy-1⁺ and Thy-1⁻ fibroblasts, suggesting a prominent role for the Thy-1⁻ phenotype in lung physiology.[24] The discordant proliferative response between Thy-1$^+$ and Thy-1$^-$ cells may be due to higher levels of IL-RA levels in Thy-1$^+$ fibroblasts. Although we have found equivalent levels of secreted IL-1 RA from mouse lung fibroblasts,[45] it remains possible that rat lung Thy-1$^-$ fibroblasts express different levels of IL-1 RA.

Human and rat Thy-1$^-$ lung fibroblasts proliferate in response to connective tissue growth factor (CTGF), an important fibroblast activator that works in concert with TGF-β to promote fibrogenesis.[46] Not only do Thy-1$^-$ fibroblasts respond to CTGF with increased proliferation, but they secrete higher levels of CTGF compared to Thy-1$^+$ cells.[47] Collectively, these data suggest that Thy-1-null fibroblasts may be of importance in the development and progression of fibrotic lung disease through their enhanced proliferative and synthetic capacities. Thus, the increased mitogenic response of this fibroblast subset may condition Thy-1$^-$ cells to participate in pro-fibrotic responses by becoming activated in the presence of chronic inflammation.[24]

The cytoskeletal organization and migratory potential of fibroblasts, important in wound healing and fibroproliferation, may also be linked to the level of Thy-1 expression on the cell surface. Barker et al[48] recently reported that rat lung fibroblasts exhibited a differential migratory potential. Thy-1$^-$ fibroblasts migrated more efficiently into wounds in vitro than Thy-1$^+$ cells, which displayed elongated focal adhesions and actin stress fibers. Increasing the level of Thy-1 in these Thy-1-null fibroblasts, using a mouse Thy-1 expression vector, resulted in fibroblasts that failed to migrate in the wound region, similar to Thy-1$^+$ fibroblasts.[48] Thy-1 expression and migratory potential correlated positively with inhibition of Src family kinase, which has been linked to cytoskeletal organization and migration.[48]

The regulation of focal adhesion assembly in Thy-1$^+$ fibroblasts is dependant on thrombospondin-1 (TSP-1), a protein that mediates the early phases of wound healing by inducing the disassembly of focal adhesions, thereby facilitating fibroblast migration.[49] When treated with TSP-1 or the hep I peptide of TSP-1, only Thy-1$^+$ fibroblasts underwent focal adhesion disassembly, an effect mediated through hep-1-induced activation of Src family kinases.[49] These finding highlight the importance of Thy-1 in regulating fibroblasts in normal and pathological processes.

Summary and Conclusions

It has become increasingly clear that fibroblasts are capable of providing more than structural support. They maintain a balance between proliferation, migration and the synthesis of immune mediators. Dysregulated inflammation and fibroproliferation underlies many disorders such as idiopathic pulmonary fibrosis. Thy-1-based separation of fibroblast subsets has demonstrated that fibroblast heterogeneity plays an important role in the progression of inflammation and appropriate resolution of wound healing. The production of cytokines by fibroblasts deficient in Thy-1 and upregulation of MHC II molecules and conversely, the generation of prostaglandins and upregulation of CD40 by Thy-1-expressing cells in many tissues highlights the contribution of both fibroblast subsets to the overall inflammatory cascade. However, tissue and species-specific differences in cytokine and prostaglandin production in fibroblast subsets highlight the concept of fibroblast heterogeneity.

Although an endogenous ligand for Thy-1 remains to be identified, recent reports suggest that Thy-1 levels directly correlate with fibroblast proliferative and migratory potential. By using a mouse Thy-1 expression vector, Barker and colleagues[48,49] have shown that Thy-1 is essential for proper fibroblast migration. Further, the assembly of focal adhesions by thrombospondin-1 in Thy-1$^+$ fibroblasts occurred through Src kinase activity.[49] This represents one of the first reports characterizing an intracellular signaling pathway mediated by Thy-1 expression on fibroblasts. Identifying the function and signaling pathway utilized by Thy-1 should prove important in better understanding the role of fibroblasts in disease. These insights may eventually lead to the tissue-specific treatment of disease such as idiopathic pulmonary fibrosis and Graves' ophthalmopathy.

Acknowledgements

This work was supported by NIH grants EY011708, EY008976, EY014564, DK63121, HL04492, ES01247, HL75432, P. Harris U.S.A./International and an American Lung Association Fellowship (C.J. Baglole).

References

1. White ES, Lazar MH, Thannickal VJ. Pathogenetic mechanisms in usual interstitial pneumonia/ idiopathic pulmonary fibrosis. J Pathol 2003; 201(3):343-354.
2. Korn JH, Thrall RS, Wilbur DC et al. Fibroblast heterogeneity: Clonal selection of fibroblasts as a model for fibrotic disease. In: Phipps RP, ed. Pulmonary fibroblast heterogeneity. Boca Raton: CRC Press Inc., 1992:119-133.
3. Kaufman J, Sime PJ, Phipps RP. Expression of CD154 (CD40 ligand) by human lung fibroblasts: Differential regulation by IFN-gamma and IL-13, and implications for fibrosis. J Immunol 2004; 172(3):1862-1871.
4. Gabbiani G. The role of contractile proteins in wound healing and fibrocontractive diseases. Methods Achiev Exp Pathol 1979; 9:187-206.
5. Breen E, Cutroneo KR. Biochemical and molecular aspects of pulmonary fibroblast heterogeneity. In: Phipps RP, ed. Pulmonary fibroblast heterogeneity. Boca Raton: CRC Press Inc., 1992:27-53.
6. Smith TJ, Sempowski GD, Berenson CS et al. Human thyroid fibroblasts exhibit a distinctive phenotype in culture: Characteristic ganglioside profile and functional CD40 expression. Endocrinology 1997; 138(12):5576-5588.
7. Smith TJ, Koumas L, Gagnon A et al. Orbital fibroblast heterogeneity may determine the clinical presentation of thyroid-associated ophthalmopathy. J Clin Endocrinol Metab 2002; 87(1):385-392.
8. Bordin S, Page RC, Narayanan AS. Heterogeneity of normal human diploid fibroblasts: Isolation and characterization of one phenotype. Science 1984; 223(4632):171-173.
9. Phipps RP, Penney DP, Keng P et al. Characterization of two major populations of lung fibroblasts: Distinguishing morphology and discordant display of Thy 1 and class II MHC. Am J Respir Cell Mol Biol 1989; 1(1):65-74.
10. McIntosh JC, Hagood JS, Richardson TL et al. Thy1 (+) and (-) lung fibrosis subpopulations in LEW and F344 rats. Eur Respir J 1994; 7(12):2131-2138.
11. Koumas L, King AE, Critchley HO et al. Fibroblast heterogeneity: Existence of functionally distinct Thy 1(+) and Thy 1(-) human female reproductive tract fibroblasts. Am J Pathol 2001; 159(3):925-935.
12. Koumas L, Phipps RP. Differential COX localization and PG release in Thy-1(+) and Thy-1(-) human female reproductive tract fibroblasts. Am J Physiol Cell Physiol 2002; 283(2):C599-608.
13. Lee WS, Jain MK, Arkonac BM et al. Thy-1, a novel marker for angiogenesis upregulated by inflammatory cytokines. Circ Res 1998; 82(8):845-851.
14. Froncek MJ, Derdak S, Felch ME et al. Cellular and molecular characterization of Thy-1+ and Thy-1- murine lung fibroblasts. In: Phipps RP, ed. Pulmonary Fibroblast Heterogeneity. Boca Raton: CRC Press Inc., 1992:135-198.
15. Zhou Y, Hagood JS, Murphy-Ullrich JE. Thy-1 expression regulates the ability of rat lung fibroblasts to activate transforming growth factor-beta in response to fibrogenic stimuli. Am J Pathol 2004; 165(2):659-669.
16. Haeryfar SM, Hoskin DW. Thy-1: More than a mouse pan-T cell marker. J Immunol 2004; 173(6):3581-3588.
17. Pont S. Thy-1: A lymphoid cell subset marker capable of delivering an activation signal to mouse T lymphocytes. Biochimie 1987; 69(4):315-320.
18. Baglole CJ, Reddy SY, Pollock SJ et al. Phenotypic characterization and isolation of primary fibroblast strains and their derivative subsets. Methods in Molecular Medicine, Fibrosis: Experimental Approaches and Protocols. Totowa: Humana; In Press.
19. Penney DP, Keng PC, Derdak S et al. Morphologic and functional characteristics of subpopulations of murine lung fibroblasts grown in vitro. Anat Rec 1992; 232(3):432-443.
20. Sempowski GD, Derdak S, Phipps RP. Interleukin-4 and interferon-gamma discordantly regulate collagen biosynthesis by functionally distinct lung fibroblast subsets. J Cell Physiol 1996; 167(2):290-296.
21. Smith TJ, Sempowski GD, Wang HS et al. Evidence for cellular heterogeneity in primary cultures of human orbital fibroblasts. J Clin Endocrinol Metab 1995; 80(9):2620-2625.
22. Borrello MA, Phipps RP. Differential Thy-1 expression by splenic fibroblasts defines functionally distinct subsets. Cell Immunol 1996; 173(2):198-206.

23. Phipps RP, Borrello MA, Blieden TM. Fibroblast heterogeneity in the periodontium and other tissues. J Periodontal Res 1997; 32(1 Pt 2):159-165.
24. Hagood JS, Mangalwadi A, Guo B et al. Concordant and discordant interleukin-1-mediated signaling in lung fibroblast thy-1 subpopulations. Am J Respir Cell Mol Biol 2002; 26(6):702-708.
25. Koumas L, Smith TJ, Feldon S et al. Thy-1 expression in human fibroblast subsets defines myofibroblastic or lipofibroblastic phenotypes. Am J Pathol 2003; 163(4):1291-1300.
26. Koumas L, Smith TJ, Phipps RP. Fibroblast subsets in the human orbit: Thy-1+ and Thy-1-subpopulations exhibit distinct phenotypes. Eur J Immunol 2002; 32(2):477-485.
27. Fries KM, Felch ME, Phipps RP. Interleukin-6 is an autocrine growth factor for murine lung fibroblast subsets. Am J Respir Cell Mol Biol 1994; 11(5):552-560.
28. Prabhakar BS, Bahn RS, Smith TJ. Current perspective on the pathogenesis of Graves' disease and ophthalmopathy. Endocr Rev 2003; 24(6):802-835.
29. Sempowski GD, Rozenblit J, Smith TJ et al. Human orbital fibroblasts are activated through CD40 to induce proinflammatory cytokine production. Am J Physiol 1998; 274(3 Pt 1):C707-714.
30. Mukaida N. Interleukin-8: An expanding universe beyond neutrophil chemotaxis and activation. Int J Hematol 2000; 72(4):391-398.
31. Zhang Y, Cao HJ, Graf B et al. CD40 engagement up-regulates cyclooxygenase-2 expression and prostaglandin E2 production in human lung fibroblasts. J Immunol 1998; 160(3):1053-1057.
32. Wang HS, Cao HJ, Winn VD et al. Leukoregulin induction of prostaglandin-endoperoxide H synthase-2 in human orbital fibroblasts. An in vitro model for connective tissue inflammation. J Biol Chem 1996; 271(37):22718-22728.
33. Kehry MR. CD40-mediated signaling in B cells. Balancing cell survival, growth, and death. J Immunol 1996; 156(7):2345-2348.
34. Noelle RJ. CD40 and its ligand in host defense. Immunity 1996; 4(5):415-419.
35. Cao HJ, Wang HS, Zhang Y et al. Activation of human orbital fibroblasts through CD40 engagement results in a dramatic induction of hyaluronan synthesis and prostaglandin endoperoxide H synthase-2 expression. Insights into potential pathogenic mechanisms of thyroid-associated ophthalmopathy. J Biol Chem 1998; 273(45):29615-29625.
36. Roy SG, Nozaki Y, Phan SH. Regulation of alpha-smooth muscle actin gene expression in myofibroblast differentiation from rat lung fibroblasts. Int J Biochem Cell Biol 2001; 33(7):723-734.
37. Liu T, Dhanasekaran SM, Jin H et al. FIZZ1 stimulation of myofibroblast differentiation. Am J Pathol 2004; 164(4):1315-1326.
38. Sime PJ, O'Reilly KM. Fibrosis of the lung and other tissues: New concepts in pathogenesis and treatment. Clin Immunol 2001; 99(3):308-319.
39. Moodley YP, Caterina P, Scaffidi AK et al. Comparison of the morphological and biochemical changes in normal human lung fibroblasts and fibroblasts derived from lungs of patients with idiopathic pulmonary fibrosis during FasL-induced apoptosis. J Pathol 2004; 202(4):486-495.
40. Bjoraker JA, Ryu JH, Edwin MK et al. Prognostic significance of histopathologic subsets in idiopathic pulmonary fibrosis. Am J Respir Crit Care Med 1998; 157(1):199-203.
41. Ramos C, Montano M, Garcia-Alvarez J et al. Fibroblasts from idiopathic pulmonary fibrosis and normal lungs differ in growth rate, apoptosis, and tissue inhibitor of metalloproteinases expression. Am J Respir Cell Mol Biol 2001; 24(5):591-598.
42. Bonner JC. Regulation of PDGF and its receptors in fibrotic diseases. Cytokine Growth Factor Rev 2004; 15(4):255-273.
43. Hagood JS, Miller PJ, Lasky JA et al. Differential expression of platelet-derived growth factor-alpha receptor by Thy-1(-) and Thy-1(+) lung fibroblasts. Am J Physiol 1999; 277(1 Pt 1):L218-224.
44. Bonner JC, Lindroos PM, Rice AB et al. Induction of PDGF receptor-alpha in rat myofibroblasts during pulmonary fibrogenesis in vivo. Am J Physiol 1998; 274(1 Pt 1):L72-80.
45. Silvera MR, Phipps RP. Synthesis of interleukin-1 receptor antagonist by Thy-1+ and Thy-1- murine lung fibroblast subsets. J Interferon Cytokine Res 1995; 15(1):63-70.
46. Watts KL, Spiteri MA. Connective tissue growth factor expression and induction by transforming growth factor-beta is abrogated by simvastatin via a Rho signaling mechanism. Am J Physiol Lung Cell Mol Physiol 2004; 287(6):L1323-L1332.
47. Hagood JS, Lasky JA, Nesbitt JE et al. Differential expression, surface binding, and response to connective tissue growth factor in lung fibroblast subpopulations. Chest 2001; 120(1 Suppl):64S-66S.
48. Barker TH, Grenett HE, MacEwen MW et al. Thy-1 regulates fibroblast focal adhesions, cytoskeletal organization and migration through modulation of p190 RhoGAP and Rho GTPase activity. Exp Cell Res 2004; 295(2):488-496.
49. Barker TH, Pallero MA, MacEwen MW et al. Thrombospondin-1-induced focal adhesion disassembly in fibroblasts requires Thy-1 surface expression, lipid raft integrity, and Src activation. J Biol Chem 2004; 279(22):23510-23516.

CHAPTER 4

Tissue Repair in Asthma:
The Origin of Airway Subepithelial Fibroblasts and Myofibroblasts

Sabrina Mattoli*

Abstract

Asthma is characterized by functional and structural alterations of the bronchial epithelium, chronic airway inflammation and remodeling of the normal bronchial architecture. Bronchial myofibroblasts are thought to play a crucial role in the pathogenesis of subepithelial fibrosis, a prominent aspect of the remodeling process. The results of the studies reviewed in this report indicate that circulating fibrocytes contribute to the bronchial myofibroblast population and may be responsible for the excessive collagen deposition below the epithelial basement membrane in asthma. More information on the mechanisms regulating the migration and differentiation of these cells in the asthmatic airways may help identify novel targets for therapeutic intervention.

Introduction

Asthma is characterized by structural and functional abnormalities of the bronchial epithelium, accumulation of inflammatory cells in the bronchial mucosa, and remodeling of the airway tissue structure.[1-9] The remodeling process has been suspected to contribute to the irreversible airflow obstruction and permanently impaired pulmonary function observed in patients with chronic asthma,[10,11] and has become a major target for the development of new anti-asthma drugs.[12]

Disruption of the epithelium integrity, accumulation of myofibroblasts below the epithelial basement membrane and subepithelial fibrosis are peculiar aspects of airway tissue remodeling in asthma.[11,13-16] In allergic asthma, these alterations may be due to an increased susceptibility of the bronchial epithelium to allergen-induced injury or to ineffective repair mechanisms during chronic allergen exposure. Tissue response is similar to that observed in chronic wounds and pathological scaring:[17] the massive apoptosis of inflammatory cells and myofibroblasts normally observed when epithelialization has been completed does not occur, as epithelial integrity is not restored, and failure to resolve the inflammatory and structural changes leads to excessive extracellular matrix deposition and deformation of the normal tissue.

In asthma, the thickened lamina reticularis beneath the airway epithelium contains abnormally high amounts of collagens I, III and V; fibronectin and tenascin.[14,18] Like in chronic wounds and other conditions where excessive extracellular matrix deposition occurs as a

*Sabrina Mattoli—Avail Biomedical Research Institute, Avail GmbH, P.O. Box 110, CH-4003 Basel, Switzerland. Email: smattoli@avail-research.com

Tissue Repair, Contraction and the Myofibroblast, edited by Christine Chaponnier, Alexis Desmoulière and Giulio Gabbiani. ©2006 Landes Bioscience and Springer Science+Business Media.

consequence of impaired wound healing, most of these proteins are thought to be produced by the numerous myofibroblasts present in the subepithelial zone,[11,14] particularly collagen I, collagen III and fibronectin. In order to find a suitable target to prevent or inhibit the remodeling process in asthma, the origin of bronchial myofibroblasts and the mechanisms involved in the accumulation of those cells need to be elucidated.

The main objective of this report is to review the data indicating that circulating fibrocytes contribute to the myofibroblast population in injured tissues, may function as precursors of bronchial myofibroblasts in asthma and may play a crucial role in the genesis of subepithelial fibrosis.

Phenotypic and Functional Characteristics of Circulating Fibrocytes

Mature fibrocytes represent a unique population of cells that express fibroblast products, such as collagen I, in conjunction with the hematopoietic stem cell antigen CD34, the leukocyte common antigen CD45 and the marker of the myeloid lineage cells CD13.[19-23] They originate from CD13+/CD45+/CD34+/collagen I− precursors present in the CD14+ fraction of peripheral blood mononuclear cells, which become CD14−/collagen I+ during the maturation process.[19,20] The phenotypic characteristics of these precursors have been poorly investigated, but their differentiation into mature fibrocytes may be up-regulated by interaction with activated T lymphocytes.[19,20]

Mature fibrocytes normally comprise less than 1% of the circulating pool of nonred cells. They do not constitutively express the myofibroblast marker α-smooth muscle actin (SMA),[19,22] but acquire the myofibroblast phenotype under in vitro stimulation with fibrogenic cytokines that are produced in exaggerated quantities in the airways of asthmatic patients, such as TGF-β_1[24] and endothelin-1.[25] When cultured in presence of these cytokines, fibrocytes develop bundles of actin microfilaments indicative of a contractile phenotype and show the other ultrastructural characteristics of fibroblasts undergoing differentiation into myofibroblasts.[22] Exposure to TGF-β_1 also increases collagen I immunoreactivity in cultured fibrocytes[19] and markedly enhances the release of collagen I, collagen III and fibronectin from these cells.[22] The differentiation of circulating fibrocytes into fibroblasts and myofibroblasts in vitro is associated with a down-regulation of the expression of the surface antigens CD34,[22,23,26] CD45[23,26] and CD13.[23] Human and murine circulating fibrocytes are quite similar in terms of phenotypic characteristics and response to TGF-β_1 in vitro.[22,23]

Phenotypic Characteristics and Bone Marrow Origin of Tissue Fibrocytes

A recent study[23] was addressed to investigate the phenotypic characteristics and bone marrow origin of fibrocytes in the wounded skin of female BALB/c mice that had received a male whole bone marrow transplant after total body irradiation (sex-mismatched bone marrow chimera mice). At 4 and 7 days post-wounding, numerous fibrocytes could be isolated from digested fragments of wounded tissue by Percoll density gradient centrifugation and immunomagnetic depletion of contaminating cells. More than 95% of the isolated cells expressed CD13 in conjunction with collagen I, as assessed by double-staining with fluorescent antibodies against CD13 and collagen I and analysis of stained cells by fluorescence-activated cell sorting. At an early stage of the healing process, a substantial proportion of these cells (mean % ± SE: 33.8% ± 8.5% from 4 to 5 experiments) also expressed the myofibroblast marker α-SMA, further increasing (to 58.7% ± 9.4%) at day 7 post-wounding. These data indicated that circulating fibrocytes rapidly enter a phase of differentiation into myofibroblasts once they have migrated at the site of tissue injury. During the differentiation process, the proportion of fibrocytes expressing CD45 and CD34 progressively decreased: on the average, more than 20% of the isolated fibrocytes lose CD45 and CD34 expression between day 4 and day 7 post-wounding.

In the same study, the bone marrow origin of fibrocytes isolated from the wounded tissue of female mice recipients of male bone marrow grafts was also examined by fluorescence in situ hybridization with a mouse Y chromosome paint probe. The Y chromosome could be identified in 94.3% or more of the fibrocytes present in the wounded tissue at day 4 post-wounding, indicating that the vast majority of these cells originated from a bone marrow-derived precursor.

Previous studies in which the origin of fibrocytes was evaluated,[21,26] provided conflicting and inconclusive data. Bucala and colleagues[21] were unable to demonstrate the bone marrow origin of fibrocytes in sex-mismatched bone marrow chimera mice, while another group could develop fibrocyte-like cells from cultured bone marrow cells of C57Bl/6 mice.[26] However, the first study was conducted in female mice that had received a male bone marrow transplant following only 800 rads of total body irradiation, which may not allow a successful ablation of the bone marrow cell population in the female recipients and effective transplantation of the bone marrow cells from male donors.

Identification of Circulating Fibrocytes as Precursors of Bronchial Myofibroblasts in Asthma

Schmidt and colleagues[22] identified cells expressing both CD34 and procollagen I mRNA in the bronchial mucosa of patients with chronic allergic asthma. The number of these cells markedly increased during an exacerbation of the disease induced by inhalation of the allergen to which patients were sensitized. At 24 hours after allergen exposure, a substantial proportion of the CD34+/procollagen I mRNA+ cells also expressed α-SMA and localized to areas of collagen deposition below the epithelial basement membrane, in a zone where myofibroblasts accumulate in chronic asthma.[11,14] The phenotype of these cells strongly suggested that they were fibrocytes migrated from the peripheral blood, which were differentiating into collagen-producing myofibroblasts at the tissue site.

This possibility was explored by using an animal model of allergic asthma that recapitulates most of the inflammatory and structural alterations of the human disease, including the accumulation of eosinophils within and below the airway epithelium and the thickening of the subepithelial zone, with deposition of collagen and other extracellular matrix proteins.[22] In this model, systemically immunized BALB/c mice were challenged with an aerosolized solution of 2.5% ovalbumin (OVA) in phosphate buffered saline (PBS) in a whole body inhalation chamber for 20 minutes, 3 times a week, at intervals of 24 hours, over a period of 8 weeks. Control mice were exposed to the OVA vehicle alone (PBS). During repeated OVA exposures, sensitized mice showed a progressive increase in the number of CD34+/procollagen I mRNA+ cells in the airway wall in comparison with control animals exposed to phosphate buffered saline (PBS). In the airway wall of animals exposed to the allergen for 6 to 8 weeks there were also numerous cells expressing the CD34 antigen in conjunction with α-SMA. By comparison with adjacent tissue sections double-labeled with antibodies against CD34 and procollagen I, it was possible to estimate that, on the average, 44.9% of the CD34+/procollagen I+ cells also expressed α-SMA at 8 weeks of repeated exposure to OVA. At the same time point, the CD34+/α-SMA+ cells represented about 31% of all cells expressing α-SMA (excluding vessels and smooth muscle cells).

Figure 1, shows the time-course of fibrocyte accumulation in the airway wall of mice chronically exposed to OVA in relation to the kinetics of two key events: the production of TGF-β_1 by airway resident cells (Fig. 1A) and the deposition of collagen I below the epithelial basement membrane (Fig. 1B). In the epithelium and subepithelial area of mice chronically exposed to OVA there was a marked increase in the number of cells showing TGF-β_1 immunoreactivity in comparison with mice chronically exposed to PBS for similar periods of time (Fig. 1A). The peak of TGF-β_1 immunoreactivity was observed between 6 and 8 weeks of chronic exposure to OVA (Fig. 1A), when many of the CD34+/procollagen I+ cells also expressed α-SMA (Fig. 1C and D). These findings suggested that the CD34+/procollagen I+/α-SMA+ cells were fibrocytes undergoing differentiation into myofibroblasts under the effect of the TGF-β_1 produced in excess by epithelial cells and inflammatory cells. Interestingly, in mice chronically exposed to OVA

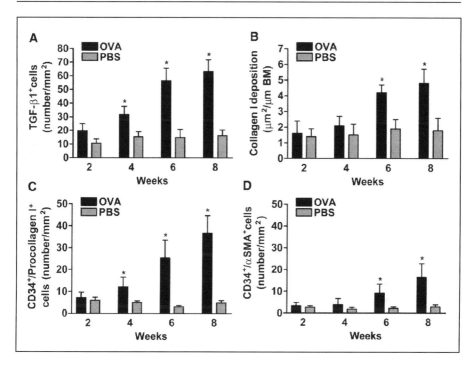

Figure 1. Kinetics of increased TGF-β_1 expression (A) and subepithelial collagen I deposition (B) in relation to the time-course of fibrocyte accumulation (C) and differentiation into myofibroblasts (D) in the airway wall of mice chronically exposed to OVA or PBS. TGF-β_1 expression was evaluated by staining tissue sections with a rabbit antibody against murine TGF-β_1. Stained cells in the epithelium and/or in the subepithelial area were counted in a zone 100 μm deep along the entire length of the epithelial basement membrane (BM) and expressed as number of cells in that zone per unit length (1 mm) of BM. Subepithelial collagen I deposition was also examined by immunohistochemical methods, using a rabbit anti-mouse collagen I antibody and an enzyme-linked indirect immunoperoxidase method. The area of collagen I staining (square μm) per length of epithelial BM (μm) was quantified by using an image analysis system. Fibrocytes coexpressing CD34 and procollagen I or CD34 and α-SMA were identified by double-staining adjacent tissue sections with an antibody against murine CD34 and an antibody against murine procollagen I or with the anti-CD34 and an antibody against intracellular α-SMA. Cell count was expressed as number of double-stained cells per square mm of airway tissue. All data are reported as the mean ± SE (n = 5-7 mice exposed to OVA or PBS at each time point). *p < 0.05 in comparison with mice exposed to PBS.

both the increase in TGF-β_1 immunoreactivity and the increase in the number of CD34$^+$/ procollagen I$^+$/α-SMA$^+$ cells occurred in concomitance with the excessive deposition of collagen I below the epithelial basement membrane (Fig. 1A-C). In view of this correlation, and considering that TGF-β_1 markedly enhances collagen release in cultured CD34$^+$/collagen I$^+$ fibrocytes acquiring α-SMA expression,[22] it is reasonable to think that the fibrocytes undergoing differentiation into myofibroblasts in the subepithelial area were an important source of that collagen and contributed to the development of subepithelial fibrosis.

Mechanisms Potentially Involved in the Recruitment of Fibrocytes into the Airways in Asthma

By tracking labeled circulating fibrocytes in the mouse model of allergic asthma, Schmidt and colleagues[22] provided direct evidence that in mice chronically exposed to OVA fibrocytes are recruited into the airway wall from the peripheral blood and localize to areas of ongoing

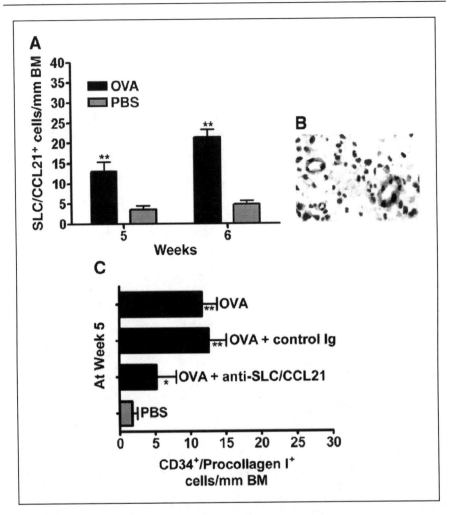

Figure 2. A) Increased expression of SLC/CCL21 in the airway wall of mice exposed to OVA for 5 and 6 weeks in comparison with control mice exposed to PBS for the same periods of time. Tissue sections were stained with an antibody against CCL21. Stained cells were counted in a zone 100 μm deep along the entire length of the epithelial basement membrane (BM) and expressed as number of positive cells in that zone per unit length (1 mm) of BM. B) Representative microphotographs showing immunoreactivity for SLC/CCL21 in the tissue section from a mice exposed to OVA for 6 weeks. Original magnification: x 200. C) Inhibition of fibrocyte accumulation by in vivo neutralization of SLC/CCL21. A 30-μg doses of either anti-CCL21 or control immunoglobulin (Ig) was injected via the tail vein 1 hour before the last OVA exposure at 5 weeks of repeated exposures. At 24 hours following that OVA exposure the number of cells coexpressing CD34 and procollagen I was assessed in the airway wall by double staining with an antibody against murine CD34 and an antibody against murine procollagen I. Stained cells were counted in a zone 100 μm deep along the entire length of the epithelial BM and expressed as number of positive cells in that zone per unit length (1 mm) of BM. All data are reported as the mean ± SE (n = 6-7 mice exposed to OVA or PBS at each time point). **p < 0.01 in comparison with mice exposed to PBS; *p < 0.025 in comparison with mice exposed to OVA that had received neither the anti-CCL21 nor the control Ig, and p < 0.05 in comparison with mice exposed to PBS.

collagen deposition below the airway epithelium. The labeled fibrocytes that accumulated in the airways had a phenotype different from the phenotype of labeled fibrocytes present in the peripheral blood. They expressed α-SMA while labeled fibrocytes present in the peripheral blood did not, and showed increased collagen I immunoreactivity. The expression of the CD34 antigen on labeled fibrocytes isolated from the airway wall was reduced in comparison with the labeled cells recovered from the peripheral blood. Taken together, these results indicated that the labeled fibrocytes recruited from the circulation into the airways had entered into a phase of differentiation into myofibroblasts at the tissue site.

The same investigators also evaluated some of the mechanisms potentially involved in the recruitment of circulating fibrocytes into the airways. The secondary lymphoid cytokine (SLC), also known as CCL21 in mice, has been shown to attract human and murine fibrocytes via the CCR7 receptor.[19] In the airway wall of mice chronically exposed to OVA for 5 to 6 weeks, there was an increased expression of SLC/CCL21 compared with the airway wall of mice exposed to PBS for the same periods of time (Fig. 2A). The sources of this chemokine were predominantly endothelial cells (Fig. 2B), but some SLC/CCL21 immunoreactivity was also observed in the epithelium and in the inflammatory infiltrate below the epithelium. The intravenous administration of a 30-μg dose of a neutralizing antibody against mouse SLC/CCL21 1 hour before the last exposure to OVA at 5 weeks significantly reduced the number of CD34[+]/procollagen I[+] cells present in the airway wall 24 hours following that OVA exposure, while the administration of a control antibody had no significant effect (Fig. 2C). However, the number of CD34[+]/procollagen I[+] fibrocytes was still significantly higher in the airway wall of OVA-challenged mice than in the airway wall of control mice (Fig. 2C). The administration of higher doses of the antibody against SLC/CCL21 did not change the results. These findings indicated that SLC/CCL21 may represent one of the factors contributing to the recruitment of circulating fibrocytes into the airways during subepithelial fibrogenesis elicited by chronic allergen exposure in BALB/c mice and that it is not the only factor involved. Another possible candidate could be the CXC chemokine stromal cell-derived factor-1 (SDF-1, also known as CXCL12), as murine and human fibrocytes seem to express the CXCR4 receptor for this chemoattractant.[19,26] However, there are conflicting data about the ability of fibrocytes to migrate in response to this chemokine in vitro and in vivo,[19,26] and more studies are needed to elucidate the role of SDF-1/CXCL12 in the recruitment of fibrocytes at the tissue sites during wound healing and in airway remodeling.

Conclusions

The data reviewed in this report indicate that circulating fibrocytes contribute to the bronchial myofibroblast population in asthma and may be involved in the genesis of subepithelial fibrosis through the release of excessive amounts of collagen and other extracellular matrix proteins below the bronchial epithelium. One of the chemokines that may induce the migration of fibrocytes from the circulation into the airways is SLC/CCL21. Once migrated at the tissue site, circulating fibrocytes may differentiate into myofibroblasts under the effect of fibrogenic cytokines produced by epithelial cells and inflammatory cells, particularly TGF-β$_1$.

It should be noted that the data discussed above do not exclude the possibility that other, TGF-β$_1$ insensitive, bone-marrow derived mesenchymal cell progenitors[27] or epithelial cells undergoing epithelial-mesenchymal transition as a result of epithelial injury[28] might also be involved in the development of subepithelial fibrosis in asthma. Thus, future studies should be addressed to investigate whether multiple cell types contribute to subepithelial fibrogenesis in this disease. Nonetheless, on the basis of the data currently available, it is reasonable to anticipate that a better understanding of the factors regulating the migration and differentiation of fibrocytes in asthmatic airways may help identify suitable targets for the development of new anti-asthma drugs.

References

1. Naylor B. The shedding of the mucosa of the bronchial tree in asthma. Thorax 1962; 17:69-72.
2. Montefort S, Djukanovic R, Holgate S et al. Ciliated cell damage in the bronchial epithelium in asthmatics and nonasthmatics. Clin Exp Allergy 1993; 23:185-189.
3. Mattoli S. Allergen-induced generation of mediators in the mucosa. Environ Health Perspect 2001; 109(Suppl 4):553-557.
4. Bucchieri F, Puddicombe SM, Lordan JL et al. Asthmatic bronchial epithelium is more susceptible to oxidant-induced apoptosis. Am J Respir Cell Mol Biol 2002; 27:179-185.
5. Bayram H, Rusznak C, Khair OA et al. Effect of ozone and nitrogen dioxide on the permeability of bronchial epithelial cells in cultures of nonasthmatic and asthmatic subjects. Clin Exp Allergy 2002; 32:1285-1292.
6. Laitinen LA, Heino M, Laitinen A et al. Damage of the airway epithelium and bronchial reactivity in patients with asthma. Am Rev Respir Dis 1985; 131:599-606.
7. Kay AB. Asthma and inflammation. J Allergy Clin Immunol 1991; 87:893-910.
8. Bousquet J, Chanez P, Lacoste JY et al. Asthma: A disease remodeling the airways. Allergy 1992; 47:3-11.
9. Elias JA, Zhu Z, Chupp G et al. Airway remodeling in asthma. J Clin Invest 1999; 104:1001-1006.
10. Fish JE, Peters SP. Airway remodeling and persistent airway obstruction in asthma. J Allergy Clin Immunol 1999; 104:509-516.
11. Gabbrielli S, Di Lollo S, Stanflin N et al. Myofibroblasts and elastic and collagen fiber hyperplasia in the bronchial mucosa: A possible basis for the progressive irreversibility of airflow obstruction in asthma. Pathologica 1994; 86:157-160.
12. Stewart AG, Tomlinson PR, Wilson J. Airway wall remodeling in asthma: A novel target for the development of anti-asthma drugs. Trend Pharmacol Sci 1993; 14:275-279.
13. Sobonya RE. Quantitative structural alterations in long-standing allergic asthma. Am Rev Respir Dis 1984; 130:289-292.
14. Brewster CEP, Howarth PH, Djukanovic R et al. Myofibroblasts and subepithelial fibrosis in bronchial asthma. Am J Respir Cell Mol Biol 1990; 3:507-511.
15. Hoshino M, Nakamura Y, Sim JJ. Expression of growth factors and remodeling of the airway wall in asthma. Thorax 1998; 53:21-27.
16. Cokugras H, Akcakaya N, Seckin S et al. Ultrastructural examination of bronchial biopsy specimens from children with moderate asthma. Thorax 2001; 56:25-29.
17. Gabbiani G. The myofibroblast in wound healing and fibrocontractive diseases. J Pathol 2003; 200:500-503.
18. Roche WR, Beasley R, Williams JH et al. Subepithelial fibrosis in the bronchi of asthmatics. Lancet 1989; 1(8637):520-524.
19. Abe R, Donnelly SC, Peng T et al. Peripheral blood fibrocytes: Differentiation pathway and migration to wound sites. J Immunol 2001; 166:7556-7562.
20. Yang L, Scott PG, Giuffre J et al. Peripheral blood fibrocytes from burn patients: Identification and quantification of fibrocytes in adherent cells cultured from peripheral blood mononuclear cells. Lab Invest 2002; 82:1183-1192.
21. Bucala R, Spiegel LA, Chesney J et al. Circulating fibrocytes define a new leukocyte subpopulation that mediates tissue repair. Mol Med 1994; 1:71-81.
22. Schmidt M, Sun G, Stacey MA et al. Identification of circulating fibrocytes as precursors of bronchial myofibroblasts in asthma. J Immunol 2003; 171:380-389.
23. Mori L, Bellini A, Stacey MA et al. Fibrocytes contribute to the myofibroblast population in wounded skin and originate from the bone marrow. Exp Cell Res 2005; 304:81-90.
24. Redington A, Madden J, Frew A et al. Transforming growth factor β1 in asthma: Measurement in bronchoalveolar lavage fluid. Am J Respir Crit Care Med 1997; 156:642-647.
25. Mattoli S, Soloperto M, Marini M et al. Levels of endothelin in the bronchoalveolar lavage fluid of patients with symptomatic asthma and reversible airflow obstruction. J Allergy Clin Immunol 1991; 88:376-384.
26. Phillips RJ, Burdick MD, Hong K et al. Circulating fibrocytes traffic to the lungs in response to CXCL 12 and mediate fibrosis. J Clin Invest 2004; 114:438-446.
27. Hashimoto N, Jin H, Chensue SW et al. Bone marrow-derived progenitor cells in pulmonary fibrosis. J Clin Invest 2004; 113:243-252.
28. Kalluri R, Neilson EG. Epithelial-mesenchymal transition and its implications for fibrosis. J Clin Invest 2003; 112:1776-1784.

CHAPTER 5

Experimental Models to Study the Origin and Role of Myofibroblasts in Renal Fibrosis

Michael Zeisberg, Mary A. Soubasakos and Raghu Kalluri*

Abstract

Most of the present knowledge on the pathomechanism of renal fibrosis is based on experimental studies with laboratory animals. Today, a variety of genetic and inducible animal models that mimic primary causes of human disease, such as diabetes mellitus, glomerulonephritis or lupus erythematodes are available. However, only few of these models progress consistently to interstitial fibrosis in the kidney involving interestitial fibrosis, tubular atrophy and glomerulosclerosis, common features of renal fibrogenesis. In this chapter, three different mouse models of human kidney disease are described highlighting their utility to study pathways leading to renal fibrosis.

The most common diseases that cause end stage renal failure (ESRF) differ significantly in their underlying primary pathomechanisms.[1,2] Glomerulonephritis due to primary glomerular inflammation, metabolic diseases such as diabetes mellitus, cystic nephropathies such as polycystic kidney disease, interstitial nephritis due to primary interstitial inflammation, and vasculopathies are among the leading causes of end stage renal failure.[1] Despite the diversity of primary patho-mechanisms associated with these different kidney diseases, they all lead to an indistinguishable scarred/fibrotic kidney.[1] The observation that chronic renal failure seems to possess common patterns independent of the underlying disease, has resulted in the speculation for the existence of a common pathogenic pathway leading to ESRF associated with fibrosis.[1,3]

COL4A3-Deficient Mice

Mice with a targeted disruption of the COL4A3 gene, which encodes for the α3 chain of type IV collagen (COL4A3), were originally generated as a model for Alport syndrome.[4,5] These mice lack normal composition of α3(IV), α4(IV) and α5(IV) type IV collagen chains in the glomerular basement membrane and display hearing defects and progressive renal disease.[4,6] The renal disease is initially characterized by splitting of the GBM (similar to the human disease), but then leads to crescentic glomerulonephritis and severe renal fibrosis.[4,7] Depending on their genetic background, these mice develop ESRD between 14 weeks (129Sv background) and 32 weeks (C57BL/6 background).[4,5] The progression of renal fibrosis in these mice stems from a defined genetic lesion, which makes interpretation of newly generated data transparent. Additionally, the progression of fibrosis within a particular genetic

*Corresponding Author: Raghu Kalluri—Harvard Medical School, Center for Matrix Biology Department of Medicine, Dana 514, Beth Israel Deaconess Medical Center, 330 Brookline Avenue Boston, Massachusetts 02215, U.S.A. Email: rkalluri@bidmc.harvard.edu

Tissue Repair, Contraction and the Myofibroblast,
edited by Christine Chaponnier, Alexis Desmoulière and Giulio Gabbiani.
©2006 Landes Bioscience and Springer Science+Business Media.

background is remarkably consistent without any further manipulations. Therefore, these mice represent a reproducible genetic mouse model, which is an excellent tool to study the progression of fibrosis in the kidney.[8]

The COL4A3-deficient mice on a 129Sv background have a normal phenotype until four weeks of age.[5,7,9] At this age the foot processes of podocytes are unaltered, the urine contains normal levels of protein, which indicates an intact glomerular filtration barrier and no pathologic alterations in the tubulointerstitium can be detected.[7] After eight weeks of age, significant splitting of the GBM is observed, associated with podocyte effacement and increased urine protein levels, indicating a malfunction within the glomerular filtration barrier. At this stage, tubulointerstitial fibrosis, the common pathway associated with several chronic nephropathies, is observed.[4,5,10] Interstitial fibrosis at this stage is indicated by a widening of the interstitial space, associated with an increase in the number of interstitial fibroblasts.[1,3,10,11] Fibrosis is furthermore characterized by an increased remodeling of the interstitial ECM, resulting in excessive deposition of fibrillar extracellular matrix (ECM).[1,11,12]

Monitoring of Disease

Progression of renal disease in COL4A3 deficient mice is reflected by proteinuria and decreased excretory renal function. Analysis of urinary protein excretion reveals the onset and progression of disease in these mice, while occurrence of hematuria is inconsistent in these mice. Blood urea nitrogen levels and serum creatinine levels increase significantly, starting between 6-8 weeks of age in these mice.

Nephrotoxic Serum Nephritis

Currently, glomerulonephritis is the underlying cause of about 10% of ESRD cases in the United States. Nephrotoxic Serum Nephritis (NTN) is a commonly used model for anti-glomerular basement membrane (anti-GBM) disease or in situ immune-complex glomerulonephritis, which features a primary glomerular insult that can then lead to tubulointerstitial injury and subsequent fibrosis.[13] NTN is characterized by two distinct disease phases: an early heterologous phase due to linear anti-GBM antibody deposition in the glomerulus that resembles anti-GBM disease and a subsequent autologous immune response against the planted antibodies resulting in an in situ immune complex formation that resembles, e.g., post infectious glomerulonephritis.[13] Preimmunization with rabbit/ sheep IgG in complete Freund's adjuvant prior to administration of a subnephritic dose of nephrotoxic serum (NTS) results in an augmentation of the autologous phase, leading to nephrotic range proteinuria and interstitial fibrosis within 3-6 weeks.[13,14]

NTN has been successfully utilized as a model for fibrosis using CD-1 mice, C57BL/6 mice and various gene knockout mice. NTS is classically raised in sheep or rabbits against crude preparations of rat glomeruli and hence contains polyclonal antibodies against multiple GBM, endothelial, mesangial and podocyte cell wall antigens. In order to induce fibrosis in CD1 mice, these mice are preimmunized by subcutaneous injection of normal sheep IgG followed by intravenous NTS injection. Fibrotic lesions (interstitial collagen deposition, increased fibroblasts, EMT) appear after 1-2 weeks of NTN. Severe tubulointerstitial fibrosis is observed between 3-6 weeks, and reaches ESRF after 4-6 weeks.

Progression of Disease

12 h after the first injection of NTS, protein casts are present in renal tubules, coinciding with the onset of proteinuria. After 24 h, leukocytes are evident within the interstitium and by day 3 prominent perivascular infiltrates are observed. At this point the glomeruli are clearly hypercellular with focal areas of necrosis. On day 5, some glomeruli contain crescents, and interstitial and perivascular infiltrates are also pronounced.[14] On day 7 about two thirds of glomeruli exhibit prominent cellular crescents and marked abnormalities are present within the tubules and interstitium.[14] These include severely dilated tubules with

flattened or denuded epithelia and expansion of the interstitium due to edema, interstitial infiltrates, and the onset of fibrosis. On day 14, the fibrotic process is more diffuse through-out the interstitium.[14] Severe tubulointerstitial fibrosis then progresses to ESRD between weeks 2 and 6.

Monitoring of Disease

Progression of NTN is reflected by proteinuria and is decreased excretory renal function. Mice injected with NTS are significantly proteinuric by day 3 and become progressively more so during the course of disease.[14] Blood urea nitrogen levels rise rapidly during the acute phase of immunologic injury, but then return to baseline values by day 10. However, concomitant with the development of progressive interstitial and glomerular fibrosis, blood urea nitrogen levels rise progressively from day 14 until the death of these mice.[14] In our experience, mice will appear normal for a long time and will only appear sick only after the disease has led to end-stage renal failure. At this stage mice lose weight rapidly and appear lethargic. Some mice develop edema due to the loss of proteins in their urine.

Unilateral Ureteral Obstruction

Unilateral Ureteral Obstruction (UUO) is a model resembling human obstructive neph-ropathy, which represents an inducible model of interstitial fibrosis secondary to a nonimmune insult. UUO is induced by ligation of an ureter of one kidney, while the contra-lateral kid-ney serves as a control. In order to obtain reliable obstruction usually two ligatures, about 5 mm apart, are placed in the upper two thirds of the ureter. An additional feature of this model is that regeneration of established lesions occurs, if the obstruction is relieved within the first three days of disease. In order to allow relief of the obstruction in a second surgery, the ureter can be protected during obstruction with a small piece of polyethylene tubing. Interstitial fibrosis, indicated by the widening of the interstitial space, associated with inter-stitial deposition of type IV collagen and tubular cell apoptosis can be observed as early as three days after obstruction.[15] If the ureteral ligation is maintained, within two weeks severe fibrosis is noticed.

Surgical Procedure

In order to obtain reliable obstruction of the ureter, two ligatures, about 5 mm apart, are placed in the upper two thirds of the ureter of the left kidney. After the ligation of the abdomi-nal cavity, muscles and skin are closed layer by layer with 6-0 nylon and 6-0 absorbable (for muscles) sutures. The duration of the whole procedure takes about 5-10 min. The sutures should be removed 10 days after the surgical procedure.

Which Is the Best Model to Use?

We have used each of these mouse models to study the progression of fibrosis in the kidney (Table 1). Each of these models consistently results in severe tubulointerstitial fibro-sis. In our experience the genetic model of COL4A3 deficient mice is a reliable, consistent model of fibrosis. This model is associated with slow progression rate, and thus might also reflect the fibrotic process in humans with kidney diseases. However, this model is currently limited to only few mouse strains, which differ substantially in their rate of progression for reasons unknown at the moment. The model of UUO is also quite consistent, it has been induced in various genetic backgrounds and it yields results in a relatively short period of time. Furthermore, this model is widely used and thus well-characterized, which allows a comparison of the newly obtained results with numerous previously published studies. How-ever, the relevance of this model to human disease is unknown, as obstruction of the ureter is less likely to lead to fibrosis in humans. Furthermore, the underlying pathomechanisms in this model differs substantially from the underlying pathomechanisms in most renal diseases associated with fibrosis.

Table 1. Summary of advantages and disadvantages of mouse models of renal fibrosis

Model	Pros	Cons
COL4A3 KO	Genetic Model, no intervention required	Strain dependency
	Consistent progression to fibrosis	
UUO	Reliable induction of fibrosis	Unclear relation human disease
	No strain dependency	
	Extensive characterization	
NTS	Crude model of crescentic glomerulonephritis	Variability in disease progression
	Fast induction of fibrosis	Strain dependency

Nephrotoxic serum nephritis, results in renal fibrosis due to a primary insult in the glomerulus, which is representative of a majority of underlying diseases leading to renal fibrosis. Progression of disease towards fibrosis occurs with 4-6 weeks, which represents an intermediate time course as compared to UUO and COL4A3-deficient mice. Even though this model is easily inducible, its utility can be limited due to substantial variation in disease progression. However, this model has been successfully used in several treatment studies, which attempted to test the therapeutic potential of novel drugs.

Histopathology and Morphometric Analysis

In order to compare the disease progression between different mouse models and different treatment protocols, a standardized procedure that evaluates interstitial fibrosis, tubular atrophy and glomerulosclerosis by morphometric analysis is required. In our laboratory we have established a standardized algorism, which is applied to all our renal fibrosis studies (Fig. 1). We routinely stain paraffin-embedded kidney sections with periodic acid Schiff (PAS), Hematoxylin Eosin (H&E) and Masson Trichrome staining for histopathological analysis. In order to quantify pathologic lesions, we determine the interstitial volume (as an indicator of interstitial fibrosis), the tubular atrophy index, glomerulosclerosis and glomerular cellularity (as parameters for glomerular injury).[8]

Interstitial volume is determined by a point-counting technique on tissue sections stained by the Masson's Trichrome method and it is expressed as the mean percentage of grid points lying within the interstitial area within five fields of renal cortex. A 1-mm^2 graded ocular grid is viewed at ~200 x magnification delineating each of these fields. Five randomly selected cortical areas, which include glomeruli, are evaluated for each animal. In order to distinguish between acute inflammation and chronic fibrosis, we evaluate the interstitial area containing blue-stained fibrous material. In order to quantify tubular atrophy, we determine a "tubular atrophy index".[8] Kidney sections are stained with periodic acid Schiff (PAS) for assessment of tubular basement membranes, and atrophic tubules are identified by their thickened and sometimes duplicated basement membranes. The number of atrophic tubules per field at ~400 x magnification is counted, and we analyze routinely about 10-20 fields per kidney cortex section. The number of atrophic tubules is then expressed as percentage of all tubules. In order to

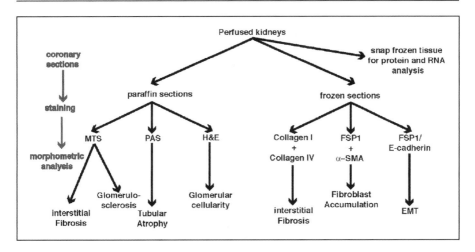

Figure 1. Standardized analysis of renal fibrosis.

quantify glomerulosclerosis, we use a semi-quantitative grading system, in which glomerulosclerotic lesions are scored for severity (0-4, 0 = normal, 4 = maximum severity) as described by Raij.[16] Additionally, we count all nuclei in 50 glomeruli of similar size in each kidney to estimate glomerular cellularity.

In order to further characterize the fibrotic lesions in a kidney, we routinely perform immunohistochemistry on frozen sections. We stain sections with antibodies to type I collagen (to assess interstitial collagen deposition), type IV collagen (to assess disruption of tubular basement membrane and ectopic expression in the interstitium), fibroblast specific protein1 (FSP1, to determine to accumulation of fibroblasts), α-smooth muscle actin (α-SMA, to assess fibroblast activation) and FSP1/E-cadherin double staining (to evaluate the occurrence of EMT).

References

1. Remuzzi G, Bertani T. Pathophysiology of progressive nephropathies. N Engl J Med 1998; 339:1448-1456.
2. Pastan S, Bailey J. Dialysis therapy. N Engl J Med 1998; 338:1428-1437.
3. Zeisberg M, Strutz F, Muller GA. Renal fibrosis: An update. Curr Opin Nephrol Hypertens 2001; 10:315-320.
4. Cosgrove D, Meehan DT, Grunkemeyer JA et al. Collagen COL4A3 knockout: A mouse model for autosomal Alport syndrome. Genes Dev 1996; 10:2981-2992.
5. Miner JH, Sanes JR. Molecular and functional defects in kidneys of mice lacking collagen alpha 3(IV): Implications for Alport syndrome. J Cell Biol 1996; 135:1403-1413.
6. Kalluri R, Cosgrove D. Assembly of type IV collagen. Insights from alpha3(IV) collagen-deficient mice. J Biol Chem 2000; 275:12719-12724.
7. Hamano Y, Grunkemeyer JA, Sudhakar A et al. Determinants of vascular permeability in the kidney glomerulus. J Biol Chem 2002; 30:30.
8. Zeisberg M, Bottiglio C, Kumar N et al. Bone morphogenic protein-7 inhibits progression of chronic renal fibrosis associated with two genetic mouse models. Am J Physiol Renal Physiol 2003; 285:F1060-1067.
9. Andrews KL, Betsuyaku T, Rogers S et al. Gelatinase B (MMP-9) is not essential in the normal kidney and does not influence progression of renal disease in a mouse model of Alport syndrome. Am J Pathol 2000; 157:303-311.
10. Cosgrove D, Rodgers K, Meehan D et al. Integrin alpha1beta1 and transforming growth factor-beta1 play distinct roles in alport glomerular pathogenesis and serve as dual targets for metabolic therapy. Am J Pathol 2000; 157:1649-1659.
11. Eddy AA. Molecular insights into renal interstitial fibrosis. J Am Soc Nephrol 1996; 7:2495-2508.
12. Border WA, Noble NA. Targeting TGF-beta for treatment of disease. Nat Med 1995; 1:1000-1001.

13. Lloyd CM, Minto AW, Dorf ME et al. Rantes and monocyte chemoattractant protein-1 (MCP-1) play an important role in the inflammatory phase of crescentic nephritis, but only MCP-1 is involved in crescent formation and interstitial fibrosis. J Exp Med 1997; 185:1371-1380.
14. Zeisberg M, Hanai J, Sugimoto H et al. BMP-7 counteracts TGF-beta1-induced epithelial-to-mesenchymal transition and reverses chronic renal injury. Nat Med 2003; 9:964-968.
15. Iwano M, Plieth D, Danoff TM et al. Evidence that fibroblasts derive from epithelium during tissue fibrosis. J Clin Invest 2002; 110:341-350.
16. Raij L, Azar S, Keane W. Mesangial immune injury, hypertension, and progressive glomerular damage in Dahl rats. Kidney Int 1984; 26:137-143.

Epithelial to Mesenchymal Transition of Mesothelial Cells as a Mechanism Responsible for Peritoneal Membrane Failure in Peritoneal Dialysis Patients

Abelardo Aguilera, Luiz S. Aroeira, Marta Ramírez-Huesca,
José A. Jiménez-Heffernan, Rafael Selgas and Manuel López-Cabrera*

Abstract

Peritoneal dialysis (PD) is an alternative to hemodialysis for the treatment of end-stage renal disease and is based on the use of the peritoneum as a semi-permeable membrane for water and solutes. Peritoneal membrane fibrosis (or sclerosis) is one of the most frequent complications of PD that includes a wide spectrum of peritoneal structural changes, ranging from mild inflammation to severe sclerosing peritonitis and encapsulating-sclerosing peritonitis. In parallel with fibrosis, the peritoneum shows a progressive increase of capillary number (angiogenesis) and vasculopathy, which are involved in increased small solute transport across the peritoneal membrane and ultrafiltration failure. Local production of vascular endothelial growth factor (VEGF) during PD appears to play a central role in the processes leading to peritoneal angiogenesis and functional decline. The most important factors of the PD solutions responsible of peritoneal deterioration are glucose and glucose degradation products, which stimulate transforming growth factor-β (TGF-β) and VEGF production by mesothelial cells (MC). TGF-β is a potent pro-fibrotic factor and inducer of epithelial-mesenchymal transition (EMT) of the MC.

This review discusses the mechanism implicated in peritoneal structural alteration and points to EMT of MC as protagonist and starter of peritoneal membrane injury through the increase of submesothelial fibroblast population. Possible mechanisms of regulation and new targets for inhibition of EMT or its deleterious effects are proposed.

Introduction

Peritoneal dialysis (PD) is a form of renal replacement that has increased during the last years, in parallel to its complications. Currently, prolonged survival on PD has been reached due to technological advances, prevention and early diagnosis of uremic complications. One of the most important issues in PD is the long-term preservation of the peritoneal membrane function.

*Corresponding Author: Manuel López-Cabrera—Unidad de Biología Molecular, Hospital Universitario de la Princesa, Diego de León, 62, 28006-Madrid, Spain.
Email: mlopez.hlpr@salud.madrid.org

Tissue Repair, Contraction and the Myofibroblast,
edited by Christine Chaponnier, Alexis Desmoulière and Giulio Gabbiani.
©2006 Landes Bioscience and Springer Science+Business Media.

The peritoneal membrane is lined by a monolayer of mesothelial cells (MC) that have characteristics of epithelial cells and act as a permeability barrier across which ultrafiltration and diffusion take place. Unfortunately, long-term exposure to hyperosmotic, hyperglycemic and low pH of dialysis solutions and repeated episodes of peritonitis or hemoperitoneum cause injury of the peritoneum, which progressively becomes denuded of MC and undergoes fibrosis and neovascularization.[1] Such structural alterations are considered the major cause of ultrafiltration failure.[1,2]

Two main reasons have led to PD-induced sclerosis to become a subject of active research: first, the high frequency of mild degree peritoneal sclerosis (50 and 80%),[3,4] and second the severity and poor prognosis of the so-called encapsulating-sclerosing peritonitis (ESP).[4,5] Fortunately, the frequency of ESP is low (0.5-4.3 cases per 1000 patients per year).[5,6] The severity of ESP and the lack of adequate and proved alternative therapeutic management deserve special attention. However, fibrosis is not the unique structural alteration of the peritoneal membrane induced by PD. In parallel with this alteration, the peritoneum shows a progressive increase of capillary number (angiogenesis) and vasculopathy, which is also related to type-I membrane failure, characterized by elevated transport of water and small solutes.[7] In this context, it has been proposed that local production of vascular endothelial growth factor (VEGF), a potent pro-angiogenic cytokine, during PD plays a central role in processes leading to peritoneal angiogenesis and functional decline.[8]

The patho-physiology of peritoneal functional impairment during long-term PD has remained elusive for long time. Recently, we have demonstrated that, soon after PD is initiated, peritoneal MC from dialysis effluents show a progressive loss of epithelial phenotype and acquire fibroblast-like characteristics.[9] In addition, by immunohistochemical studies of peritoneal biopsies from PD patients, we demonstrated the expression of the mesothelial markers in stromal α-smooth muscle actin (αSMA)-positive myofibroblasts,[9,10] suggesting that these cells stemmed from local conversion of MC. All these biochemical and morphological changes of the MC are reminiscent of those that take place during the EMT, also called trans-differentiation. EMT is a complex and generally reversible process that starts with the disruption of intercellular junctions and loss of apical-basolateral polarity, typical of epithelial cells, which are then transformed into fibroblast-like cells with pseudopodial protrusions and increased migratory, invasive, and fibrogenic features.[11] MC have been considered, for long time, as mere victims of the peritoneal injury during long-term PD, whereas peritoneal stromal fibroblasts have been classically considered as the main cells responsible of the structural and functional peritoneal alterations.[9]

Our most recent findings show a clear association between EMT of MC, synthesis of VEGF, secretion of extracellular matrix components (ECM) and type-I peritoneal membrane failure (our unpublished results). Nowadays, there is no treatment for the progressive thickening and angiogenesis of peritoneal membrane associated with PD. The observation that EMT of MC is a key process in the initiation of peritoneal fibrosis and angiogenesis, opens new insights for therapeutic intervention. The therapeutic treatments may be designed toward either the direct prevention of EMT of the MC or its deleterious effects such as ECM synthesis and /or VEGF production.

Peritoneal Fibrosis

Peritoneal fibrosis (or sclerosis) is a term that comprises a wide spectrum of peritoneal structural alterations, ranging from mild inflammation to severe sclerosing peritonitis and its most complicated manifestation ESP.[3-5] Simple sclerosis (SS), an intermediate stage of peritoneal fibrosis, is the most common peritoneal lesion found in the patients after few months on PD, and could represent the initial phase of sclerosing peritonitis (SP).[2] Rubin et al[2] described a normal thickness of the peritoneum of 20 μm, but after a few months on PD could reach up to 40 μm (SS). The SP is a progressive sclerosis that is characterized by a dramatic thickening of

Table 1. Sequential peritoneal structural changes associated with medium-long-term PD (2, 6-8)

- Loss of mesothelial cells microvilli and intercellular junctions
- Round profile of the mesothelial cells nuclei
- Increase in nuclear heterochromatin
- Hyperplasia with degenerative changes of rough endoplasmic reticulum
- Pyknosis or swelling of mitochondria
- Increase in cellular turnover
- Loss of cellular regeneration capacity
- The peritoneum is completely denuded of mesothelial cells
- Finally, the thickness of the peritoneum as a subjacent phenomenon

the peritoneum (up to 4000 μm) and is accompanied by inflammatory infiltrates, calcification, neo-vascularization and dilatation of blood and lymphatics vessels, being the thickening the most commonly used pathological criterion for differential diagnosis.[12-14] In some instances, granulated tissue is observed immersed in exudates containing fibrin and giant cells, probably reflecting chronic inflammation.

Peritoneal fibrosis consists of the accumulation of ECM proteins (collagen I, III, V, VI, fibronectin, tenascin) in the interstitium, with augmented number of fibroblasts, some of them with myofibroblastic features, and mononuclear cell infiltration. In the basement membrane there is usually accumulation of collagen IV and laminin and proteoglycans, polysaccharides and glycoproteins are also present extracellularly.[3-5] Table 1, summarize the sequential structural changes in peritoneal membrane during PD.

There are multiple factors involved in the sclerotic process of the peritoneum which may be dialysis-dependent and dialysis-independent:

1. Peritonitis is one of the most commonly invoked pathogenic factors for SP.[5,6,13] Some etiological agents have been identified including the bacteria *Staphylococcus aureus, Pseudomona sp.* and *Haemophilus influenza.* These pathogens promote conversion of fibrinogen by coagulase to a molecular form of fibrin particularly resistant to breakdown by plasmin.[15] The mechanism by which peritonitis promotes progression to SP may start by the denudation of the mesothelium,[4] which facilitates the peritoneal damage by the bioincompatible compounds from PD solutions,[16] increases peritoneal permeability to glucose, and favours non-enzymatic glycosylation of submesothelial structural proteins. Another important issue is the loss of fibrinolytic capacity by MC which increases the ratio production/reabsorption of fibrin. Furthermore, peritonitis is associated with the increased intraperitoneal expression of transforming growth factor-β (TGF-β) and other cytokines and growth factors that may accelerate the fibrotic process of the peritoneum.

2. Time on PD: some authors,[12-14] but not others,[5,6] have found a relationship between months on PD and the incidence of SP. The main factor appears to be the prolonged exposure to glucose from PD solution,[16] which is able to stimulate TGF-β and fibroblast growth factor (FGF) productions by MC.[17-20] In addition, we have observed a correlation between the time on PD and the progression EMT of the MC.[9]

3. Poor biocompatibility of dialysis fluids: high glucose concentration, low pH and lactate buffer in current PD solutions, are all factors that have been implicated in peritoneal fibrosis.[16,21,22] These compounds have been associated to diminished production of phospholipids by MC, impaired phagocytosis capacity of macrophages,[23] decreased activation of neutrophils[24] and lymphocytes,[25] and direct toxicity of fibroblasts.[26] Although the PD fluid components are risk factors for SP,[24,25] it is not always possible to identify the triggering agents for the progression of SP.

Chlorhexidine and povidone iodine, employed to sterilize PD connections and preparation of surgeries, have been also implicated in the progression of SP.[27,28] The peritoneal catheter as well as bags and tubes for dialysis are other risk factors that may cause reactive fibrosis.[5]

4. As peritoneal dialysis-independent factors, the administration of β-blockers has been associated to SP development by a decrease in surfactant release by MC, similar to that observed in type-II pneumocytes.[15] Stegmayr[29] and Krediet[30] have demonstrated the association between β-blockers and reduced ultrafiltration related to a decrease in portal vein pressure and an increase in lymphatic drainage from the peritoneal cavity.

On the other hand, evidence for genetic predisposition to SP has also been proposed.[4]

Role of TGF-β in the Pathogenesis of Peritoneal Fibrosis

TGF-β is a growth factor which has been implicated as causal agent in fibrosis of different tissues and organs such as skin, lung, eye, central nervous system and diseases like acute mesangial proliferative glomerulonephritis.[18]

TGF-β synthesis may be stimulated by glucose from PD solutions, and acute or chronic infections, via peritoneal leukocyte-derived factors (Fig. 1). In this regard, TGF-β has been found to be up-regulated in peritoneal inflammatory processes and its over-expression has been correlated to worse PD outcomes.[31] TGF-β exists in tissues, generally, as a latent and inactive form, bound to the latency-associated peptide (LAP), and it is activated through proteolytic cleavage by thrombospondin, plasmin, cathepsin D, furin, and glycosidases when exposed to

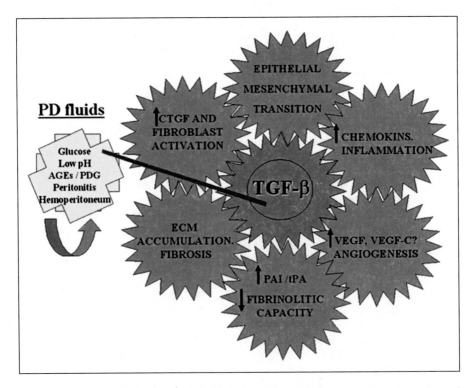

Figure 1. TGF-β is considered the master molecule in peritoneal injury during PD. Glucose, advanced glycation end products (AGE), glucose degradation products (GDP), low pH from PD fluids and peritonitis stimulate TGF-β synthesis, which in turn triggers the healing processes that ultimately lead to tissue fibrosis and angiogenesis.

Table 2. Implication of TGF-β in peritoneal fibrosis

- Activates quiescent fibroblasts into myofibroblasts (33)
- Increases fibronectin production by fibroblast and MC (34, 35).
- Induces the expression of connective tissue growth factor (CTGF) by MC (36).
- Induces EMT of MC (9, 63).
- Stimulates the synthesis of PAI, the natural inhibitor of tPA, contributing to the generation of an anti-fibrinolitic environment (37).
- Increases matrix synthesis and inhibits matrix degradation by decreasing the ratio MMPs/TIMPs (18).

acid pH as occurs with PD solutions.[32] However, this important activation step has received little attention in vivo.

TGF-β is considered the master molecule in the genesis of peritoneal fibrosis, because it plays a central role in the triggering and perpetuation of many wound healing processes[33-37] (Fig. 1). Table 2, summarizes the effects of TGF-β in fibrosis.

The relevance of TGF-β in peritoneal fibrosis has been demonstrated in an in vivo mouse model, in which TGF-β gene was transduced into the peritoneal cavity with an adenovirus vector, reproducing the structural and functional alterations observed in PD patients.[8]

The TGF-β family includes TGF-β1, TGF-β2, TGF-β3, activins, and bone morphogenic proteins (BMP). These cytokines interact with the subfamily of TGF-β receptors type II, and then recruit and activate the subfamily of TGF-β receptors type I, which triggers the cellular signalling pathways through its serine/threonine kinase activity.[18,38] TGF-β is considered one of the most important inducer of EMT of different epithelial cells, including the MC, both in vitro and in vivo.[39] Four different intracellular signal pathways are triggered upon engagement of TGF-β to its receptors (Fig. 2), being the most important the Smads cascade. TGF-β-receptors I phosphorylates Smad 2 and 3 inducing their association with the common partner Smad 4, and then they translocate into the nucleus, where they control the expression of TGF-β-responsive genes, such as that encoding integrin-linked kinase (ILK).[38] The activation of up-regulated ILK by β1 integrins results in strong phosphorylation of Akt and glycogen synthase kinase-3 (GSK-3).[40]

Phosporylated Akt triggers NF-κB activation,[41] which in turn induces the expression of Smad7,[42] an inhibitory Smad molecule that interferes with the phosporylation of Smad2/3, and of snail, a key regulator of EMT. The transcription factor snail regulates EMT by inhibiting the expression of E-cadherin,[43,44] and by inducing growth arrest and survival, which confer selective advantage to migrating trans-differentiated cells[45] (Fig. 2). The phosphorylation of GSK-3 by ILK results in its inhibition and subsequently stabilization of β-catenin, released from the adherens junctions, and of AP-1.[46] Stabilized β-catenin, in conjunction with Lef-1/Tcf, may per-se induce EMT,[47] and AP-1 activates MMP-9 expression inducing invasion of the ECM[48] (Fig. 2).

One of the main Smad-independent signaling cascades triggered by TGF-β receptor I ligation, include the RhoA-p160ROCK pathway which regulates cytoskeleton remodeling and cellular migration/invasion[49] as well as NF-κB activation.[50] In addition, RhoA induces the expression of α-SMA in a ROCK-independent manner[49] (Fig. 2).

Another signal transduction stimulated by TGF-β is the H-Ras/Raf/ERK pathway which is also necessary for the induction of snail expression and EMT.[51,52] In this context, it has been described that TGF-β and fibroblast growth factor (FGF), a potent inducer of the H-Ras/Raf/ERK signaling, cooperate in the triggering of EMT.[53]

Finally, TGF-β may induce apoptosis via activation of the mitogen activated protein kinase P38.[54] However, this pro-apoptotic effect of TGF-β may be counteracted by NF-κB activation (Fig. 2).

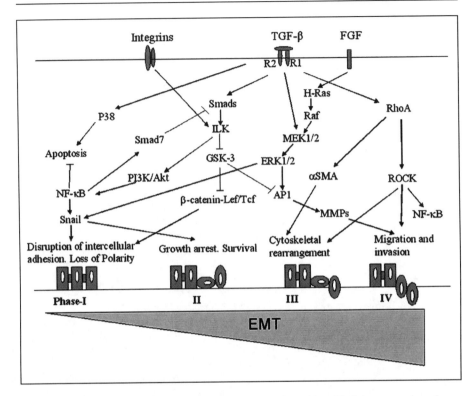

Figure 2. Intracellular signal cascades induced by TGF-β implicated in epithelial to mesenchymal transition (EMT). Four different intracellular signal pathways are triggered upon engagement of TGF-β to its receptors, being the most important the Smads cascade. Other Smad-independent signaling cascades triggered by TGF-β receptor I ligation, include the RhoA-p160ROCK, the H-Ras/Raf/ERK and the mitogen activated protein kinase P38 pathways.

Implication of Epithelial-Mesenchymal Transition of MC in Peritoneal Fibrosis

EMT is a physiologic process necessary for tissue repair in normal condition and if uncontrolled may promote a fibrotic response.[55] EMT is controlled by the interplay of a number of extracellular and intracellular molecules and signaling pathways.[56] Extracellular factors involved in EMT include TGF-β, epidermal growth factor (EGF), interleukin-1 (IL-1), angiotensin II, advanced glycation end products (AGEs), connective tissue growth factor (CTGF), and FGF. Interestingly, the presence of MMP2 or collagen-I alone promote EMT, while collagen-IV suppresses it.[57] Regarding the intracellular processes involved in EMT, TGF-β-induced signals appear to be the most relevant in the establishment and maintenance of the mesenchymal state.[56]

In PD, several factors may induce EMT, these include peritonitis through the synthesis of pro-inflammatory cytokines such as IL-1; hemoperitoneum by activating the plasminogen-fibrin pathway; low pH through activation of latent TGF-β, elevated AGEs in peritoneal membrane;[18,31,33,56,58,59] and glucose from PD fluids by direct stimulation of TGF-β secretion.[60]

The presence of a high population of fibroblasts is always representative of fibrosis, but their origin is debated. The classic concept suggests that fibroblasts are simply residual embryonic mesenchymal cells that remained in the different tissue after organogenesis. However, the current view argues that fibroblasts may arise from local conversion of epithelial cells by EMT or from

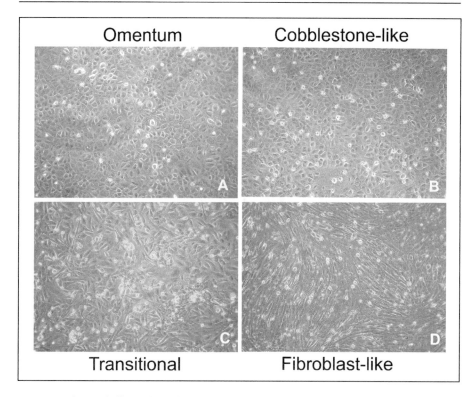

Figure 3. Cultures of effluent-derived mesothelial cells (MC). Phase-contrast microscopy showing different morphology of MC: omentum-derived cells are the control of normal MC, and cobblestone-like, transitional and fribroblast-like MC were obtained from PD effluents.

CD34 [+] cells (fibrocytes) of the bloodstream after being recruited from bone marrow.[61,62] In the case of PD-related fibrosis, we have demonstrated that, soon after PD is initiated, MC obtained from the PD effluents show a progressive loss of epithelial morphology and acquire a fibroblast-like phenotype[9] (Fig. 3). These transformed MC progressively lose the expression of epithelial markers such as E-cadherin and cytokeratins, up-regulate the expression of snail, fibronectin and collagen I and acquire migratory capacity.[9] In immunohistochemical studies of peritoneal biopsies from PD patients, we demonstrated the expression of mesothelial markers in stromal spindle-like cells, suggesting that they stemmed from local conversion of MC[9,10] (Fig. 4). The transformed MC express the myofibroblast molecule α-smooth muscle actin (α-SMA)[10] (Fig. 5). We and others have shown that this myofibroblastic conversion of MC can be induced in vitro with various stimuli.[6,63] Our findings, suggest that new fibroblasts may arise mainly from local conversion of MC by EMT during the repair responses of the peritoneal tissue induced by PD. In contrast, we did not observe a significant contribution of CD34[+] cells from bone marrow to the submesothelial fibroblast population in the fibrotic peritoneal tissue. In the case of renal fibrosis models, it has been shown that 36% of new fibroblasts derive from EMT and 15% from bone marrow. The rest comes from local proliferation of resident fibroblasts.[64]

Role of Epithelial-Mesenchymal Transition of MC in Neovascularization and Peritoneal Transport Disorders

There is emerging evidence that expansion of the peritoneal vasculature and augmented vessel permeability are important determinants of increased small solute transport across the

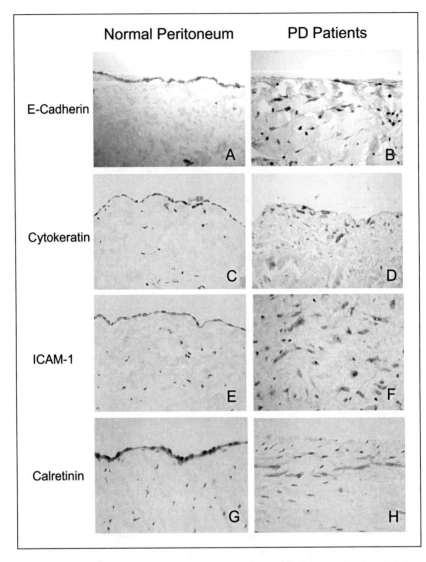

Figure 4. Expression of mesothelial markers in stromal spindle-like cells. Immunohistochemical studies of peritoneal biopsies show the Immunoexpression of E-cadherin, cytokeratins, ICAM-1, and calretinin in cells with fibroblastic morphology in PD patients, suggesting that they stemmed from local conversion of MC.

peritoneal membrane and ultrafiltration failure.[65] Vascular endothelial growth factor (VEGF) is a potent pro-angiogenic cytokine involved in endothelial cell proliferation and vascular permeability. It has been proposed that local production of VEGF during PD plays a central role in processes leading to peritoneal angiogenesis and functional decline.[66] The main source of VEGF in PD patients as well as the mechanisms implicated in VEGF up-regulation during PD remain unclear. Previous studies have shown that MC from omentum have the capacity to produce VEGF in vitro in response to a variety of stimuli such as glucose degradation products (GDPs), AGEs, and TGF-β.[59,66,67] Furthermore, it has been reported that effluent-derived

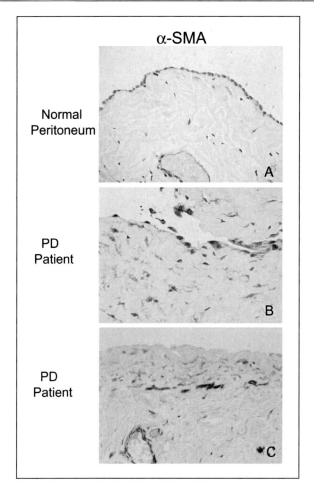

Figure 5. Expression of α-smooth muscle actin (α-SMA) in normal peritoneum and in peritoneal samples from PD patients. Except for vessels there is no expression of α-SMA in normal peritoneum. In PD patients, immunoexpression of α-SMA is evident in a small subset of mesothelial cells with preserved morphology and in submesothelial cells with fibroblast-like morphology.

MC cells produce spontaneously different levels of VEGF ex vivo, but the reason for these different VEGF production abilities were not established.[68] Recently, we have found that the mechanism underlying VEGF up-regulation in MC is the EMT of these cells, which is induced by PD itself and by the episodes of peritonitis or hemoperitoneum (our unpublished results).

The association between peritoneal fibrosis and ultrafiltration failure and high solute transport across the peritoneal membrane is well known. In addition to fibrosis, an increased number of capillary vessels is also related to type-I membrane failure, which is characterized by increased transport of water and small solutes.[65,69] However, the relation between peritoneal transport features and EMT has not been established so far. Our most recent data show a clear association between EMT, synthesis of VEGF and ECM proteins and type-I peritoneal membrane failure (our unpublished results).

Therapeutic Intervention on EMT

Taking into account that EMT is an important target to prevent fibrosis and angiogenesis of peritoneal tissue, therapeutic approaches may be addressed to prevent either EMT of the MC or its deleterious effects such as ECM synthesis and/or VEGF production. In regard of the prevention of EMT of MC, it has been described that endogenous factors such as hepatocyte growth factor (HGF) and BMP-7 are able to inhibit and reverse EMT and renal fibrosis in animal models.[70,71] In addition, it has been shown that EMT induced in vitro by TGF-β, can be reversed by HGF and BMP-7 by promoting the recovering of E-cadherin expression and down-regulating the expression of mesenchymal markers such as α-SMA, vimentin and fibronectin.[70,71] The mechanism by which HGF interferes with EMT is by inducing the expression of the Smad co-repressor SnoN, which interacts with the translocated Smad-complex and block the expression of Smad-dependent genes, including ILK. The mechanism underlying BMP-7 counteraction of EMT is by activation a subset of Smad proteins that antagonize with those (Smad2/3) activated by TGF-β.[71]

Other strategies that would open new avenues of therapeutic intervention to prevent or reverse EMT of MC may include the inhibition of ILK, RhoA-ROCK or Akt-mediated signaling cascades.[49,54,72] In this context, the administration of the ROCK inhibitor Y-27632 resulted in suppression of α-SMA expression and renal interstitial fibrosis in a mouse model of ureteral obstruction.[72]

Based on the concept, that EMT, fibrosis and angiogenesis may be part of the same process of peritoneal membrane failure, another therapeutic approach may be desigened to prevent its deleterious effects such as ECM synthesis and/or VEGF production. In this context, it has been shown that rapamycin has an anti-tumoral effect through inhibition of VEGF synthesis by the tumor cells.[73] Thus, this drug may also have a potential therapeutic use for preventing peritoneal vessel expansion and ultrafiltration failure. Our recent data supported this notion and demonstrated a reduction of VEGF synthesis capacity of trans-differentiated MC in vitro. Importantly, the synthetic estrogen tamoxifen has demonstrated to have a clear protective and anti-fibrotic effect in PD patients with high risk of suffering encapsulating-sclerosing peritonitis.[74] In addition, we have found a decrease of ECM protein synthesis by transdifferentiated MC treated with different doses of this estrogen (our unpublished data).

Others substances with expected effects on peritoneal fibrosis include: pentoxifylline, dipyridamole, and emodin, which have direct inhibitory effects on ECM protein synthesis or on TGF-β expression and activity in MC.[75-77] The statin simvastatin is another candidate molecule for therapeutic treatment of peritoneal fibrosis because it exhibits a fibrinolytic activity by inducing tPA synthesis and inhibiting PAI-1 expression.[78] Finally, the synthetic glucocorticoid prednisolone may also have a clinical potential since it inhibits glucose-induced basic fibroblast growth factor expression by MC, and hence prevents submesothelial fibroblasts proliferation.[79] Table 3, shows several drugs and therapeutic strategies to reverse EMT or prevent it deleterious effects.[80-88]

Conclusion

In conclusion, recent findings suggest that in the peritoneum new fibroblast-like cells arise from local conversion of MC by EMT during the repair responses that take place in long-term PD. The trans-differentiated MC may retain a permanent mesenchymal state, as long as initiating stimuli persist, and may contribute to PD-induced fibrosis and angiogenesis and peritoneal membrane failure (Fig. 6). EMT appears as the central point in the pathogenesis of peritoneal damage associated to PD. Thus, future therapeutic approaches for preventing or reversing peritoneal functional decline in PD patients should be designated in order to prevent or reverse EMT of MC or its pernicious effects.

Table 3. Drugs and strategies to reverse EMT or its deleterious effects

Agents	Mechanisms	References
Anti-fibrotic agents		
• Tamoxifen	Inhibition ECM production Smads pathway inhibition	74*
• (AcSDKP) Tetrapeptide	TGF-β inhibition	80
• Liver growth factor	Inhibition of ECM production	81
• Pentoxifylline	Inhibition of ECM production	75
• Dipyrodamole	Inhibition of TGF-β production	76
• Emodin	Inhibition of ECM production	77
• Simvastatin	Increase fibrinolytic activity	78
Anti-angiogenic agents		
• Anecortave acetate	Inhibits VEGF production	82
• Pegaptanib	Inhibits VEGF-VEGFR-binding	83
• Roxatidine	Decrease VEGF expression	84
• anti-VEGFRII	Blocking receptor VEGFRII	85
• Rapamycin	Inhibition of VEGF production	73*
• TNP-470	Decreases VEGF expression Inhibition of ECM production	86
• FTY720	Decreases VEGF expression	87
Inhibition of EMT		
• BMP-7	TGF-β/Smad pathway inhibition	71
• ILK-inhibitors (HGF)	ILK inhibition	54, 56, 70, 88
• Y-27632	Rho/ROCK-inhibitor	72
• Antioxidant agent	NF-κB inhibition	51

*Our-unpublished data. (AcSDKP): N-Acetyl-seryl-aspartyl-lysyl-proline

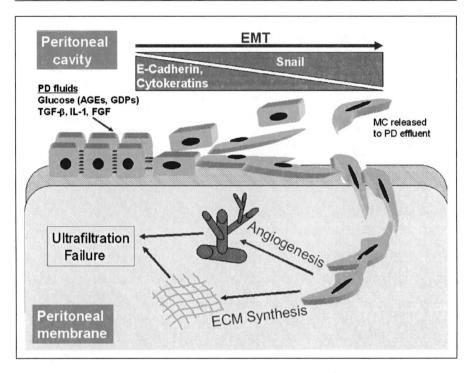

Figure 6. In the peritoneum new fibroblast-like cells arise from local conversion of MC by EMT during PD. The trans-differentiated MC may retain a permanent mesenchymal state, as long as initiating stimuli persist, and may contribute to PD-induced fibrosis and angiogenesis and peritoneal membrane failure.

Acknowledgements

This work was supported by grants F.I.S. 03/0599 and CAM 08.8/0009.1/2003 to R. Selgas, and by grants SAF-2004-07855 and C03/02 to M. López-Cabrera. This work was also partially supported by a research award from Fundación I.C.O. to M. López-Cabrera.

References

1. Selgas R, Bajo MA, Del Peso G et al. Preserving the peritoneal dialysis membrane in long-term peritoneal dialysis patients. Sem Dial 1995; 8:326-332.
2. Rubin J, Herrera GA, Collins D. An autopsy study of peritoneal cavity from patients on continuous ambulatory peritoneal dialysis. Am J Kidney Dis 1991; 18:97-102.
3. Schneble F, Bonzel KE, Waldherr R et al. Peritoneal morphology in children treated by continuous ambulatory peritoneal dialysis. Pediatr Nephrol 1992; 6:542-546.
4. Di Paolo N, Garosi G. Peritoneal sclerosis. J Nephrol 1999; 12:347-361.
5. Nomoto Y, Kawaguchi Y, Kubo H et al. Sclerosing encapsulating peritonitis in patients undergoing continous ambulatory peritoneal dialysis: A report of sclerosing encapsulating peritonitis group. Am J Kidney Dis 1996; 28:420-427.
6. Rigby RJ, Hawley CM. Sclerosing peritonitis: The experience in Australia. Nephrol Dial Transplant 1998; 13:154-159.
7. Krediet RT, Lindholm B, Rippe B. Pathophysiology of peritoneal membrane failure. Perit Dial Int 2000; 20(Suppl 4):S22-S42.
8. Margetts PJ, Kolb M, Hoff CM et al. The role of angiopoietins in resolution of angiogenesis resulting from adenoviral mediated gene transfer of TGF-β1 or VEGF to the rat peritoneum. J Am Soc Nephrol 2001; 12:2029-2039.
9. Yáñez-Mo M, Lara-Pezzi E, Selgas R et al. Peritoneal dialysis induces an epithelial-mesenchymal transition of mesothelia cells. New Eng J Med 2003; 348:403-413.

10. Jimenez-Heffernan JA, Aguilera A, Aroeira LS et al. Immunohistochemical characterization of fibroblast subpopulation in normal peritoneal tissue and in peritoneal dialysis-induced fibrosis. Virchow Arch 2004; 444:247-256.

11. Thiery JP. Epithelial-mesenchymal transition in development and pathologies. Curr Opin Cell Biol 2003; 15:740-746.

12. Holland P. Sclerosing encapsulanting peritonitis in chronic ambulatory peritoneal dialysis. Clin Radiol 1990; 41:19-23.

13. Ing TS, Daugirdas JT, Gandhi VC. Peritoneal sclerosis in peritoneal dialysis patients. Am J Nephrol 1984; 4:173-176.

14. Carbonnel F, Barrie F, Beaugerie L et al. Sclerosing peritonitis. A serie of 10 cases and review of the literature. Gastroenterol Clin Biol 1995; 19:876-882.

15. Dobbie JW. Role of imbalance of intracavitary fibrin formation and removal in the pathogenesis of peritoneal lesions in CAPD. Perit Dial Int 1997; 17:121-124.

16. Holmes CJ. Biocompatibility of peritoneal dialysis solutions. Perit Dial Int 1993; 13:88-94.

17. Freser D, Wakefield L, Phillips A. Independent regulation of transforming growth factor beta 1 transcription and translation by glucose and platelet-derived growth factor. Am J Pathol 2002; 161:1039-1049.

18. Border WA, Noble NA. Transforming growth factor β in tissue fibrosis. N Engl J Med 1994; 331:1286-1292.

19. Lai KN, Lai KB, Lam CW et al. Changes of cytokines profile during peritonitis in patients on continuous ambulatory pertoneal dialysis. Am J Kidney Dis 2000; 35:644-652.

20. Teshima-Kondo S, Kondo K, Prado-Lourenco L et al. Hyperglycemia upregulates translation of the fibroblast growth factor 2 mRNA in mouse aorta via internal ribosome entry site. FASEB J 2004; 18:1583-1585.

21. Ogata S, Yorioka N, Kiribayashi K et al. Viability of, and basic fibroblast growth factor secretion by human peritoneal mesothelial cells culture with various components of perotoneal dialysis fluid. Adv Perit Dial 2003; 19:2-5.

22. Breborowicz A, Oreopoulos DG. Biocompatibility of peritoneal dialysis solutions. Am J Kidney Dis 1996; 27:738-743.

23. Davies SJ, Ogg CS, Cameron JS. Peritoneal macrophages markers and phagocytic function in patients on CAPD. Nephrol Dial Transplant 1998; 13:837-845.

24. Liberek T, Topley N, Jörres A et al. Peritoneal dialysis fluids inhibition of polymorphonuclear leukocyte respiratory burst activation is related to lower intracellular pH. Nephron 1993; 65:260-265.

25. Wieslander AP, Nordin MK, Martinson E et al. Heat-sterilized PD-fluids impair growth and inflammatory responses of culture cells lines and human leukocytes. Clin Nephrol 1993; 39:343-348.

26. Wieslander AP, Nordin MK, Kjellstrand PT et al. Toxicity of peritoneal dialysis fluids on cultured fibroblasts, L-929. Kidney Int 1991; 40:77-79.

27. Lo WK, Chan KT, Leung AC et al. Sclerosing peritonitis complicating prolonged use of chlorexidine in alcohol in the connection procedure for continuous ambulatory peritoneal dialysis. Perit Dial Int 1991; 11:166-172.

28. Keating JP, Neill M, Hill GL. Sclerosing encapsulanting peritonitis after intraperitoneal use of povidone iodine. Aust N Z J Surg 1997; 67:742-744.

29. Stegmayr BG. Beta-blockers may cause ultrafiltration failure in peritoneal dialysis patients. Perit Dial Int 1997; 17:541-545.

30. Krediet RT. Beta-blockers and ultrafiltration failure. Perit Dial Int 1997; 17:528-531.

31. Offner FA, Feichtinger H, Stadlmann S et al. Transforming growth factor beta synthesis by human peritoneal mesothelial cells. Induction by interleukin-1. J Pathol 1996; 148:1679-1688.

32. Schultz-Cherry S, Lawley J, Murphy-Ullrich JE. The type I repeats of thrombospondin 1 active latent transforming growth factor-beta. J Biol Chem 1994; 269:26783-26788.

33. Desmouliere A, Geinoz A, Gabbiani F et al. Transforming growth factor-β1 induces α-smooth muscle actin expression in granulation tissue myofibroblasts and in quiescent and growing cultured fibroblast. J Cell Biol 1993; 122:103-111.

34. Viedt C, Büugel A, Hänsch GM. Fibronectin synthesis in tubular epithelial cells: Up-regulation of the EDA splice variant by transforming growth factor β. Kidney Int 1995; 48:1810-1817.

35. Gharaee-Kermani M, Wiggins R, Wolber F et al. Fibronectin is the mayor fibroblast chemoattractant in rabbit anti-glomerular basement membrane disease. Am J Pathol 1996; 148:961-967.

36. Zarrinkalam KH, Stanley JM, Gray J et al. Connective tissue growth factor and its regulation in the peritoneal cavity of peritoneal dialysis patients. Kidney Int 2003; 64:331-338.

37. Eddy AA. Expression of genes that promote renal interstitial fibrosis in rats with proteinuria. Kidney Int 1996; 49:S49-S54.

38. Massagué J, Wotton D. Transcriptional control by the TGF-β/Smad signaling. EMBO 2000; 19:1745-1754.
39. Fan JM, Ng YY, Hill PA et al. Transforming growth factor-beta regulate tubular epithelial-myofibroblast transdifferentiation in vitro. Kidney Int 1999; 56:1455-1467.
40. Massagué J. How cells read TGF-beta signals. Nat Rev Mol Cell Biol 2000; 1:169-178.
41. Tan C, Mui A, Dedhar S. Integrin-linked kinase regulate inducible nitric oxide synthase and cyclooxygenase-2 expression in an NF-κB-dependent manner. J Biol Chem 2002; 277:3109-3116.
42. Bitzer M, Von Gersdorff G, Liang D et al. A mechanism of supression of TGF-β/SMAD signaling by NF-κB/RelA. Genes and Development 2000; 14:187-197.
43. Cano A, Perez-Moreno MA, Rodrigo I et al. The transcription factor snail controls epithelial-mesenchymal transitions by repressing E-cadherin expression. Nat Cell Biol 2000; 2:76-83.
44. Poser I, Dominguez D, de Herreros AG et al. Loss of E-cadherin expression in melanoma cells involves up-regulation of the transcriptional repressor Snail. J Biol Chem 2001; 276:24661-24666.
45. Vega S, Morales AV, Ocaña OH et al. Snail blocks the cell cycle and confers resistance to cell death. Genes and Development 2004; 18:1131-1143.
46. D'Amico M, Hulit J, Amanatullah DF et al. The integrin-linked kinase regulates the cyclin D1 gene through glycogen synthase kinase 3β and cAM-responsive element-binding protein-dependent pathways. J Biol Chem 2000; 275:32649-32657.
47. Kim K, Lu Z, Hay ED. Direct evidence for a role of beta-catenin/LEF-1 signaling pathway in induction of EMT. Cell Biol Int 2002; 26:463-476.
48. Troussard AA, Costello P, Yoganathan TN et al. The integrin linked kinase (ILK) induces an invasive phenotype via AP-1 transcription factor-dependent upregulation of matrix metalloproteinase 9 (MMP-9). Oncogene 2000; 16:5444-5452.
49. Masszi A, Di Ciano, Sirokmany G et al. Central role for Rho in TGF-beta1-induced alpha-smooth muscle actin expresion during epithelial-mesenchymal transition. Am J Physiol Renal Physiol 2003; 284:F911-F924.
50. Benitah SA, Valeron PF, Lacal JC. ROCK and nuclear factor-kappaB-dependent activation of cyclooxygenase-2 by Rho GTPases: Effects on tumor growth and therapeutic consequences. Mol Biol Cell 2003; 14:3041-3054.
51. Huber MA, Azoitei N, Baumann B et al. NF-κB is essential for epithelial-mesenchymal transition and metastasis in a model of breast cancer progression. J Clin Invest 2004; 114:569-581.
52. Barbera MJ, Puig Y, Domínguez D et al. Regulation of snail transcription during epithelial to mesenchymal transition of tumor cells. Oncogene 2004; 23:7345-7354.
53. Peinado H, Quintanilla M, Cano A. Transforming growth factor β-1 induces snail transcription factor in epithelial cell lines. J Biol Chem 2003; 278:21113-21123.
54. Li Y, Yang J, Dai C et al. Role for intergrin-linked kinase in mediating tubular epithelial to mesenchymal transition and renal interstitial fibrogenesis. J Clin Invest 2003; 112:503-516.
55. Ten Dike P, Miyazono K, Helding CH. Signaling inputs converge on nuclear effector in TGF-beta signaling. Trends Biochem Sci 2000; 25:64-70.
56. Liu Y. Epithelial to mesenchymal transition in renal fibrogenesis: Pathologic significance, molecular mechanism, and therapeutic intervention. J Am Soc Nephrol 2004; 15:1-12.
57. Bottinger EP, Bitzer M. TGF-β signaling in renal disease. J Am Soc Nephrol 2002; 13:2600-2610.
58. Rosenber ME. Peritoneal dialysis: Diabetes of the peritoneal cavity. J Lab Clin Med 1999; 134:103-104.
59. Oldfield MD, Bach LA, Forbes JM et al. Advanced glycation end products cause epithelial-myofibroblast transdifferentiation via the receptor for advanced glycation end products (RAGE). J Clin Invest 2001; 108:1853-1863.
60. Wang T, Chen YG, Ye RG et al. Effect of glucose on TGF-β 1 expression in peritoneal mesothelial cell. Adv Perit Dial 1995; 11:7-10.
61. Abe R, Donnelly SC, Peng T et al. Peripheral blood fibrocytes: Differenciation pathway and migration to wound site. J Immunol 2001; 166:7556-7562.
62. Kalluri R, Neilson EG. Epithelial-mesenchymal transition and its implications for fibrosis. J Clin Invest 2003; 112:1776-1784.
63. Yang AH, Chen JY, Lin JK. Myofibroblastic conversion of mesothelial cells. Kidney Int 2003; 63:1530-39.
64. Iwano M, Plieth D, Danoff TM et al. Evidence that fibroblast derived from epithelium during tissue fibrosis. J Clin Invest 2002; 110:341-350.
65. Margetts P, Bonniaud P. Basic mechanisms and clinical implications of peritoneal fibrosis. Perit Dial Int 2003; 23:530-541.

66. Mandl-Weber S, Cohen CD, Haslinger B et al. Vascular endothelial growth factor production and regulation in human peritoneal mesothelial cells. Kidney Int 2002; 61:570-578.
67. Witowski J, Jorres A, Korybalska K et al. Glucose degradation products in peritoneal dialysis fluid: Do they harm? Kidney Int 2003; 84(suppl 1):S148-S151.
68. Selgas R, del Peso G, Bajo MA et al. Vascular endothelial growth factor (VEGF) levels in peritoneal dialysis effluent. J Nephrol 2001; 14:270-4.
69. Pecoits-Filho R, Araujo MR, Lindholm B et al. Plasma and dialysate IL-6 and VEGF concentrations are associated with high peritoneal solute transport rate. Nephrol Dial Transplant 2002; 17:1480-6.
70. Yang J, Liu Y. Blockage of tubular epithelial to myofibroblast transition by hepatocyte growth factor prevents renal interstitial fibrosis. J Am Soc Nephrol 2002; 13:96-207.
71. Zeisberg M, Hanai JC, Sugimoto H et al. BMP-7 counteracts TGF-β1-induced epithelial-to-mesenchymal transition and reverses chronic renal injury. Nat Med 2003; 9:964-968.
72. Nagatoya K, Moriyama T, Kawada N et al. Y-27632 prevents tubulo-interstitial fibrosis in mouse kidneys with unilateral ureteral obstruction. Kidney Int 2002; 61:1684-1695.
73. Guba M, von Breitenbuch P, Steinbauer M et al. Rapamycin inhibits primary and metastatic tumor growth by antiangiogenesis: Involvement of vascular endothelial growth factor. Nat Med 2002; 8:128-35.
74. del Peso G, Bajo MA, Gil F et al. Clinical experience with tamoxifen in peritoneal fibrosing syndromes. Adv Perit Dial 2003; 19:32-35.
75. Fang CC, Yen CJ, Chen YM et al. Pentoxifylline inhibits human peritoneal mesothelail cell growth and collagen synthesis: Effects in TGF-β. Kidney Int 2000; 57:2626-2633.
76. Hung KY, Chen CT, Huang JW et al. Dipyrodamole inhibits TGF-beta-induced collagen gene expression in human peritoneal mesothelial cells. Kidney Int 2001; 60:1249-1257.
77. Chan TM, Leung JK, Tsang RC et al. Emodin ameliorates glucose-induced matrix synthesis in human peritoneal mesothelial cells. Kidney Int 2003; 64:519-533.
78. Haslinger B, kleemann R, Toet KH et al. Simvastatin suppress tissue factor expression and increase fibrinolytic activity in tumor necrosis factor-α-activated human peritoneal mesothelial cells. Kidney Int 2003; 63:2065-2074.
79. Ogata S, Yorioka N, Kohno N. Glucose and prednisolone alter basic fibroblast growth factor expression in peritoneal mesothelial cells and fibroblasts. J Am Soc Nephrol 2001; 12:2787-2796.
80. Kanasaki K, Koya D, Sugimoto T et al. N-Acetyl-seryl-aspartyl-lysyl-proline inhibits TGF-beta-mediated plasminogen activator inhibitor-1 expression via inhibition of Smad pathway in human mesangial cells. J Am Soc Nephrol 2003; 14:863-72.
81. Díaz Gil JJ, Rau C, Machin C et al. Hepatic growth induced by injection of the liver growth factor into normal rats. Growth Regul 1994; 4:113-22.
82. Penn JS, Rajaratnam VS, Collier RJ et al. The effect of an angiogenic steroid neovascularization in rat model of retinopathy of prematurity. Invest Ophthalmol Vis Sci 2001; 42:283-290.
83. Vinores SA. Technology evaluation: Pegaptanib, Eyetech/Pfizer. Curr Opin Mol Ther 2003; 5:673-679.
84. Tomita K, Izumi K, Okabe S. Raxatidine-cimetidine-induced angiogenesis inhibition suppresses growth of colum cancer implants in syngeneic mice. L Pharmacol Sci 2003; 93:321-330.
85. Li R, Xiong DS, Shao XF et al. Production of neutralizing monoclonal antibody against human vascular endothelial growth factor receptor II. Acta Pharmacol Sin 2004; 25:1292-1298.
86. Yoshio Y, Miyazaki M, Abe K et al. TNP-470, an angiogenesis inhibitor, suppresses the progression of peritoneal fibrosis in mouse experimental model. Kidney Int 2004; 66:1677-1685.
87. Sanchez T, Estrada-Hernandez T, Paik JH et al. Phosphorylation and action of the immunomodulator FTY720 inhibits vascular endothelial cell growth factor-induced vascular permeability. J Biol Chem 2003; 278:47281-47290.
88. Yoganathan TN, Costello P, Chen X et al. Integrin-linked kinase (ILK): A "hot" therapeutic target. Biochem Pharmacol 2000; 60:1115-1119.

FIZZy Alveolar Epithelial Cells Induce Myofibroblast Differentiation

Sem H. Phan*

Abstract

While some progress has been made recently in identifying potential candidate genes that may be important in pathogenesis of pulmonary fibrosis, the list certainly is not complete. Using DNA microarray analysis to analyze the lung gene expression profiles in a rat model of pulmonary fibrosis has revealed over 600 genes that were ≥ 2-fold up or down-regulated. The highest up-regulated gene is identified as FIZZ1 (Found in Inflammatory Zone), and found to be expressed primarily by lung epithelial cells and not in fibroblasts. Further analysis shows that FIZZ1 stimulated α-smooth muscle actin and type I collagen expression independently of transforming growth factor-β, suggesting its potential as an inducer of myofibroblast differentiation. IL-4 and IL-13 are found to induce FIZZ1 expression in type II alveolar epithelial cells via a STAT6 mediated mechanism, and interestingly IL-4/IL-13 or STAT6 deficient mice shows reduced lung FIZZ1 expression and fibrosis. Thus the potential role of FIZZ1 in fibrosis represents an example of epithelial-mesenchymal crosstalk of significance to the genesis of myofibroblasts in lung fibrosis.

Introduction

The de novo genesis of myofibroblasts in lung repair/fibrosis is considered a key event in the subsequent remodeling that occurs in affected tissues.[1] These myofibroblasts represent key sources of both extracellular matrix and pro-fibrotic cytokines such as transforming growth factor (TGF-β), and thus are considered to be a key cell type responsible for propagation of the fibrotic response. Fibrosis associated with the idiopathic interstitial pneumonias is commonly a progressive and untreatable disease terminating in respiratory failure and death. The presence of myofibroblasts in fibrosis in the lung has been amply documented.[1] Animal model studies, such as bleomycin-induced pulmonary fibrosis in rodents, have confirmed the de novo genesis of myofibroblasts with their distinct α-smooth muscle actin (α-SMA) expressing phenotype. Abundant studies have documented potential roles for a number of mediators, such as cytokines and chemokines, in the pathogenesis of pulmonary fibrosis, and with recent focus on the fibroblastic focus as a key diagnostic and negative prognostic indicator, intense interest has been generated vis-à-vis genesis of such a focus, and especially the myofibroblast located within it. One hypothesis is that repeated injury-instigated cross talk between epithelium and underlying mesenchyme via soluble mediators may be an important mechanism for genesis and accumulation of fibroblasts and myofibroblasts in fibroblastic foci.[2,3] Indeed expression of a number of

*Sem H. Phan—University of Michigan Medical School, Department of Pathology, Ann Arbor, Michigan 48109-0602, U.S.A. Email: shphan@umich.edu

Tissue Repair, Contraction and the Myofibroblast,
edited by Christine Chaponnier, Alexis Desmoulière and Giulio Gabbiani.
©2006 Landes Bioscience and Springer Science+Business Media.

mediators by alveolar epithelium with activating effects on fibroblasts has been reported, and many of these have been implicated in fibrosis. It is likely however that the list of mediators that may be involved in vivo in promoting the genesis of the myofibroblast and fibrosis is not complete, since suppression/neutralization of known mediators often does not suppress fibrosis completely. Traditional efforts at identifying additional potential candidate genes of such mediators have used a combination of analysis of expression of suspected genes, examining the effects of their neutralization/inhibition. These have yielded important clues for the role of some mediators in fibrosis, but cannot be used to efficiently discover the whole spectrum of candidate genes. The development and explosion in the use of DNA microarray technology however has enabled some recent advances in global analysis of gene expression, affording the option of simultaneous analysis of multiple genes in an animal model during the progression from initial injury to fibrosis. Recent successful use of this approach has identified potentially significant patterns of gene expression, including one study documenting a significant role for matrilysin in the bleomycin model.[4] In this report, the results are summarized of studies using a similar approach as a means of identifying additional novel genes that may be involved in pulmonary fibrosis, specifically with respect to genesis of the myofibroblast.

FIZZ1 Expression

The strategy uses a 10,000 element (10 K) rat cDNA microarray in an attempt to identify genes which are differentially expressed in bleomycin-injured rat lung tissue as the model evolved.[5] The animal model is induced by endotracheal injection of bleomycin into Fisher 344 rats and the lungs are then harvested at selected time points for isolation of RNA, fibroblasts and type II alveolar epithelial cells (AECs). Reverse transcribed isolated RNA is used for microarray analysis using this 10K rat chip.[6] Analysis of the kinetic expression data set reveals 5 distinct clusters, within which are identified 628 identified genes that were ≥ 2-fold down- or up-regulated at the various time points. The different clusters contain variable numbers of genes with different kinetic patterns and degrees or levels of expression.[5,7,8] Many of the genes have previously been reported to be up-regulated in this model and involved either in inflammation (e.g., cytokines/chemokines) and/or fibrosis (e.g., extracellular matrix components). Especially noteworthy is the 9 genes contained in one of the 5 clusters displaying an expression pattern of rapid up-regulation by day 7 after injury and sustained elevated expression until day 21. This cluster reveals some genes such as, neogenin, phospholipase D2, and FIZZ1, which have not previously been implicated in fibrosis. FIZZ1 stands out in this list because the amplitude of its increased expression in bleomycin-injured lungs is the greatest (25-fold increase on day 21) of all genes in all clusters examined, hence the rationale to focus in on this gene. FIZZ1 or resistin-like molecule α belongs to a novel family of secreted cysteine-rich proteins having 5 members thus far identified in the mouse.[9-13] They characteristically contained 10 cysteine residues that are spaced in unique fashion. Human homologs have been identified for some but not all of the members found in rodents. The biological activities of these molecules are uncertain and variable depending on the isoform. Although resistin has been implicated in insulin resistance, this activity remains controversial. Other studies of FIZZ1 have suggested its ability to inhibit nerve growth factor effects on neuron survival and adipocyte differentiation without affecting cell proliferation. More recently, there is also evidence that it can stimulate lung microvascular smooth muscle cell proliferation and play a role in lung development.[14,15] However its potential role in fibrosis is unknown, thus requiring further investigation.

Validation of the increased FIZZ1 expression in the bleomycin model by real time PCR confirms the microarray results, namely a >20-fold increase in lung FIZZ1 expression in this animal model with similar kinetics. A previous report shows increased FIZZ1 expression in an animal model of airway disease, which localizes to airway epithelial cells primarily.[9] This is confirmed in the bleomycin model using in situ hybridization, which reveals primary expression in AECs and bronchiolar epithelial cells.[5] Analysis of purified type II alveolar epithelial cells in culture confirms this finding as well as the much lower level of expression in cells

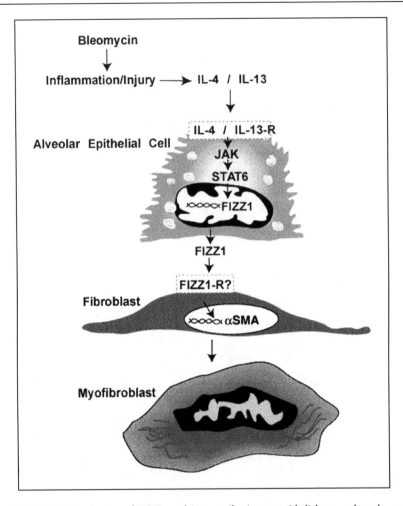

Figure 1. IL-4/IL-13 induction of FIZZ1 and its contribution to epithelial-mesenchymal crosstalk, myofibroblast differentiation and pulmonary fibrosis. In this diagram, the induction of IL-4 and IL-13 in bleomycin-induced pulmonary fibrosis is shown to induce JAK/STAT mediation of epithelial cell FIZZ1 expression. Secreted FIZZ1 can then induce adjacent fibroblasts presumably via as yet unidentified FIZZ1 binding receptors, to express α-smooth muscle actin and other genes to assume a myofibroblast phenotype. The differentiated myofibroblast can then participate in the fibrotic process by elaboration of extracellular matrix and cytokines, as well as contribute to alteration in mechanical properties of lung tissue as has been reported in fibrotic lung tissue.

isolated from uninjured control lungs. In contrast, FIZZ1 expression cannot be detected in lung fibroblasts in normal or injured lung tissue, or upon isolation and propagation in tissue culture from these tissues.[5]

Effects of FIZZ1 on Fibroblasts and Myofibroblast Differentiation

Initial attempts at evaluating the potential role of FIZZ1 examine its ability to regulate fibroblast function or phenotype. Since AECs from bleomycin-injured lungs express high levels of FIZZ1, initial studies using cocultures of these cells with fibroblasts shows increased collagen I and α-smooth muscle actin expression,[5] indicating that the FIZZ1 from AECs may have induced these effects in fibroblasts, since the AECs do not express either collagen I or this actin isoform. However fibroblast proliferation is not affected in this coculture. These findings are consistent with AEC derived FIZZ1 and/or other mediators having the ability to induce myofibroblast differentiation. This can also be observed by morphological changes in the fibroblasts consistent with myofibroblast differentiation, namely organization of actin fibers and assumption of a more spread out and adherent morphology. It is unlikely that TGF-β1 from the AECs is responsible for this activity since expression of this cytokine is not significantly elevated in AECs from injured lung. Moreover antibodies to this cytokine fail to abrogate the ability of AECs to induce myofibroblast differentiation. However the possibility of other AEC-derived mediators with such activity is not ruled out by the study using anti-TGF-β antibodies. To address this issue, rat FIZZ1 is cloned and used to create a GST-FIZZ1 fusion protein, which exhibits similar activity on isolated fibroblasts. This confirms that FIZZ1 does have the ability to induce myofibroblast differentiation and thus could play this role in vivo and promote fibrosis. The totality of the evidence indicates that highly induced FIZZ1 in AECs in this animal model can promote myofibroblast differentiation and thus may play a role in inducing the genesis of myofibroblasts in the context of pulmonary fibrosis.

Regulation of FIZZ1 Expression

In view of this potential role in myofibroblast differentiation and fibrosis, the basis for FIZZ1 induction in injured lungs would be informative in understanding the overall mechanism of how it may plays this role in pulmonary fibrosis. There is evidence that Th2 cytokines are able to do this in vitro in isolated macrophages and a bone marrow-derived cell line.[16,17] Screening a number of cytokines reveals that only the Th2 cytokines, IL-4 and IL-13 are able to stimulate FIZZ1 expression in isolated AECs in a dose-dependent manner.[18] Further analysis indicates that this is a rapid and primary response to the cytokines, and not surprisingly, this effect appears to be mediated via JAK1 and STAT6. In vivo studies using IL-4 and/or IL-13 deficient mice confirm the importance of these cytokines in induction of FIZZ1 expression in lungs of mice treated with bleomycin.[18] Lung FIZZ1 expression is significantly reduced in the respective cytokine deficient mice compared to wild type controls, and virtually absent in doubly (IL-4 and IL-13) deficient mice. Since STAT6 is implicated in IL-4/IL-13 induction of FIZZ1 expression, STAT6 deficient mice can also be used to confirm its in vivo importance in regulating FIZZ1 expression. Indeed such deficient mice show a markedly reduced ability to induce FIZZ1 in response to bleomycin-induced lung injury relative to wild type control mice, which exhibits significant bleomycin-induced increase in lung STAT6 expression. Given that lung IL-4 and IL-13 expression is elevated in bleomycin-injured lungs, these findings suggest that this induced cytokine expression may represent the responsible trigger for induction of lung FIZZ1 expression via a STAT6 dependent mechanism in this animal model.

FIZZ1 in Pulmonary Fibrosis

There is ample evidence for the importance of IL-4 and IL-13 in fibrosis in various organ systems.[19-23] With respect to pulmonary fibrosis, there is evidence that IL-4 and IL-13 may be important in the propagation of chronic fibrosis.[20,21] When IL-4/IL-13 deficient mice are evaluated for response to bleomycin, the observed diminished fibrosis relative to wild type mice correlates with the reduced lung FIZZ1 expression in the deficient animals.[18] Thus in vivo, reduced lung FIZZ1 expression affords protection from bleomycin-induced pulmonary fibrosis. Thus induction of FIZZ1 by IL-4/IL-13 via STAT6 in lung epithelial cells may promote fibrosis by promoting the genesis of myofibroblasts, making this molecule another potentially novel target for the treatment of pulmonary fibrosis. This crosstalk between epithelial cells and fibroblasts may be a key factor in the genesis of the fibroblastic focus, which is known to be a negative prognostic indicator in human pulmonary fibrosis.[24] However the direct applicability of these findings in rodents to human disease needs to await identification of a FIZZ1 homolog in humans.

Conclusion

A number of molecules are known to have the property of inducing myofibroblast differentiation in the context of tissue fibrosis, including especially TGF-β. However it is unclear in vivo as to the actual identity of the mediator or mediators that may be directly involved in genesis of myofibroblasts in fibrosis. While inhibition of TGF-β activity in vivo can suppress fibrosis and myofibroblast differentiation, this effect is mostly incomplete, suggesting additional TGF-β-independent mechanisms may also be important. The studies of FIZZ1 expression in a rodent model of pulmonary fibrosis suggest a new candidate mediator with such activity. Moreover the characteristic differential cellular expression pattern provides support for the hypothesis on the importance of epithelial-mesenchymal crosstalk in the context of repeated alveolar epithelial injury as postulated in the pathogenesis of the idiopathic form of pulmonary fibrosis or usual interstitial pneumonia. Further work is necessary to confirm this scenario in human pulmonary fibrosis, and which must await identification of the corresponding gene or gene product with similar activities in humans. Nevertheless, these new findings in rodent models have suggested an additional TGF-β-independent mechanism by which myofibroblasts could arise in lung injury as a result of epithelial-mesenchymal crosstalk, and participate in propagation of fibrosis.

Acknowledgement

This work was supported by NIH grants HL28737, HL31963, HL52285 and HL77297.

References

1. Gharaee-Kermani M, Phan SH. Role of fibroblasts and myofibroblasts in idiopathic pulmonary fibrosis. In: Lynch J, ed. Idiopathic Pulmonary Fibrosis. Vol. 20. New York: Marcel Dekker, 2003:501-563.
2. Selman M, King TE, Pardo A. Idiopathic pulmonary fibrosis: Prevailing and evolving hypotheses about its pathogenesis and implications for therapy. Ann Intern Med 2001; 134:136-151.
3. Gross TJ, Hunninghake GW. Idiopathic pulmonary fibrosis. New Engl J Med 2001; 345:517-525.
4. Zuo F, Kaminski N, Eugui E et al. Gene expression analysis reveals matrilysin as a key regulator of pulmonary fibrosis in mice and human. Proc Natl Acad Sci USA 2002; 99:6292-6297.
5. Liu T, Dhanasekaran SM, Jin H et al. Induction of FIZZ-1 expression in lung injury and fibrosis. Am J Pathol 2004; 164:1315-1326.
6. Chinnaiyan AM, Huber-Lang M, Kumar-Sinha C et al. Molecular signatures of sepsis: Multiorgan gene expression profiles of systemic inflammation. Am J Pathol 2001; 159:1199-209.
7. Romoni MF, Sebastiani P, Kohane IS. Cluster analysis of gene expression dynamics. Proc Natl Acad Sci USA 2002; 99:9121-9126.
8. Eisen MB, Spellman PT, Brown PO et al. Cluster analysis and display of genome-wide expression patterns. Proc Natl Acad Sci USA 1998; 95:14863-14868.
9. Holcomb IN, Kabakoff RC, Chan B et al. FIZZ1 a novel cysteine-rich secreted protein associated with pulmonary inflammation, defines a new gene family. The EMBO J 2000; 19:4046-4055.
10. Steppan CM, Brown EJ, Wright CM et al. A family of tissue-specific resistin-like molecules. Proc Natl Acad Sci USA 2001; 98:502-506.
11. Steppan CM, Bailey ST, Bhat S et al. The hormone resistin links obesity to diabetes. Nature 2001; 409:307-312.
12. Gerstmayer BD, Küsters S, Gebel T et al. Identification of RELMγ, a novel resistin-like molecular with a distinct expression pattern. Genomics 2003; 81:588-595.
13. Chumakov AM, Kubota T, Walter S et al. Identification of murine and human XCP1 genes as C/EBP-epsilon-dependent members of FIZZ/Resistin gene family. Oncogene 2004; 23:3414-25.
14. Teng X, Li D, Champion C et al. FIZZ1/RELMα, a novel hypoxia-induced mitogenic factor in lung with vasoconstrictive and angiogenic properties. Circ Res 2003; 92:1-5.
15. Wagner KF, Hellberg AK, Balenger S et al. Hypoxia-induced mitogenic factor has antiapoptotic action and is upregulated in the developing lung: Coexpression with hypoxia-inducible factor-2α. Am J Respir Cell Mol Biol 2004; 31:276-82.
16. Raes GP, Baetselier D, Noël W et al. Differential expression of FIZZ1 and Ym1 in alternatively versus classically activated macrophages. J Leukoc Biol 2002; 71:597-602.

17. Stütz AM, Pickart LA, Trifilieff A et al. The Th 2 cell cytokines IL-4 and IL-13 regulate found in inflammatory zone 1/resistin-like molecule α gene expression by a STAT 6 and CCAAT/ enhancer-binding protein-dependent mechanism. J Immunol 2003; 170:1789-1796.

18. Liu T, Jin H, Ullenbruch M et al. Regulation of FIZZ1 expression in bleomycin-induced lung fibrosis: Role of IL-4/IL-13 and mediation via STAT-6. J Immunol 2004; 173:3425-3431.

19. Zhu ZR, Homer J, Wang Z et al. Pulmonary expression of interleukine-13 causes inflammation, mucus hypersecretion, subepithelial fibrosis, physiologic abnormalities and eotaxin production. J Clin Invest 1999; 103:779-788.

20. Huaux FT, Liu B, McGarry M et al. Dual roles of interleukin-4 (IL-4) in lung injury and fibrosis. J Immunol 2003; 170:2083-2092.

21. Jakubzick C, Choi ES, Joshi BH et al. Therapeutic attenuation of pulmonary fibrosis via targeting of IL-4- and IL-13-responsive cells. J Immunol 2003; 171:2684-93.

22. Kolodsick JE, Toews GB, Jakubzick C et al. Protection from fluorescein isothiocyanate-induced fibrosis in IL-13-deficient, but not IL-4-deficient, mice results from impaired collagen synthesis by fibroblasts. J Immunol 2004; 172:4068-76.

23. Kaviratne M, Hesse M, Leusink M et al. IL-13 activates a mechanism of tissue fibrosis that is completely TGF-beta independent. J Immunol 2004; 173:4020-9.

24. King Jr TE, Schwarz MI, Brown K et al. Idiopathic pulmonary fibrosis: Relationship between histopathologic features and mortality. Am J Respir Crit Care Med 2001; 164:1025-32.

CHAPTER 8

Pro-Invasive Molecular Cross-Signaling between Cancer Cells and Myofibroblasts

Olivier De Wever* and Marc Mareel

Abstract

Cancer cell invasion necessitates the participation of host cells. One of the cell types that stimulates invasion of colon and other cancer cells is the myofibroblast, as evidenced from the histology of cancer and from coculture experiments. Cancer cells produce transforming growth factor-β (TGF-β) and TGF-β converts fibroblasts into pro-invasive myofibroblasts. In the in vitro system with human cancer cell lines and freshly isolated stromal cells, the pro-invasive activity of myofibroblasts is due to the combined action of Hepatocyte growth factor/scatter factor (HGF/SF) and tenascin-C, two molecules known to promote invasion in clinical tumors and their experimental surrogates. The myofibroblasts are themselves invasive and this activity is stimulated by TGF-β. N-cadherin is implicated in the invasion response of myofibroblasts. The question now is which of the multiple factors present in the tumor ecosystem is responsible for the pro-invasive switch that turns a benign tumor into a malignant one.

Host Cells Participate at Cancer Cell Invasion

Epithelial tumors consist of cancer cells and host cells. The presence of host cells is considered as a reaction against the aberrant behavior of the cancer cells, the latter being at the origin of the tumor. This host reaction was understood, originally, as a mechanical and immunological defense against the cancer cells and was described by the pathologist as desmoplasia, inflammation and neoangiogenesis with accumulation of stromal fibroblasts, leukocytes and endothelial cells respectively. Today, there is growing evidence that tumor-infiltrated host cells are recruited by the cancer cells and diverted by the latter to contribute to their malignant progression rather than to protect the host.[1] New blood and lymph vessels,[2] immunocytes and inflammatory cells[3] as well as stromal fibroblasts[4] don't inhibit but rather stimulate cancer invasion and metastasis, in line with Paget's "seed" and "soil" hypothesis.[5] Next to recruitment by cancer cells, alternative scenarios for the participation of cancer cells and host cells at tumor progression should be considered. Both cell types may undergo concomitant but independent alterations. Using serial analysis of gene expression (SAGE) during breast carcinogenesis, Allinen et al[6] found in epithelial, myoepithelial, myofibroblastic, leukocytic and endothelial cell types extensive changes in the expression of genes that encode secreted proteins and receptors. Genetic alterations, however, were found only in the epithelial cancer cells. Others have shown genetic

*Corresponding Author: Olivier De Wever—Laboratory of Experimental Cancerology, Department of Radiotherapy, University Hospital Ghent, De Pintelaan 185, B-9000 Ghent, Belgium. Email: olivier.dewever@ugent.be

Tissue Repair, Contraction and the Myofibroblast,
edited by Christine Chaponnier, Alexis Desmoulière and Giulio Gabbiani.
©2006 Landes Bioscience and Springer Science+Business Media.

mutations in stromal cells, examples being *SMAD4, LKB1, HPP1*.[7-9] Moreover, changes in the epithelial compartment might be secondary to alterations of the stroma; a phenomenon called landscaper defect.[10]

Regardless the sequence of events, we will assume, here, that there exists a continuous cross-signaling between the epithelial and the stromal compartment in normal and in pathological situations and we will discuss some of the critical alterations that lead to invasion. In view of the present symposium our host cell of interest is the myofibroblast. This name was used to describe, in experimental granulation tissue of Wistar rats, fibroblastic cells with a smooth muscle cell-like morphology, a strongly developed microfilamentous apparatus and a contractile phenotype.[11,12] The most reliable molecular myofibroblast marker is α-smooth muscle actin (α-SMA).

Myofibroblasts Stimulate Invasion

Colon cancer cells PROb, a cell line derived from a chemically induced BDIX rat tumor, produced upon subcutaneous injection into syngeneic rats ulcerating tumors that were poorly differentiated and invaded into skin and muscle with isolated cancer cells in the stroma.[13] These invasive cancers were rich in myofibroblasts, that were positive for α-SMA and localized mainly at the front of invasion. PROb cells harvested from routine cell culture, however, failed to invade when confronted in vitro with collagen, Matrigel or embryonic chick heart tissue. By contrast, freshly dissociated tumor cell suspensions, containing PROb cells and tumor-associated stromal cells, were invasive in all three in vitro assays. The role of the stromal myofibroblasts in invasion was confirmed by experiments in which PROb cells were mixed with cells of the established myofibroblast cell line DHD-FIB that was isolated from a similar BDIX rat colon tumor. The histology of invasion in vitro and in vivo gave the impression that solitary PROb cancer cells were dragged by massively invading myofibroblasts. The pro-invasive activity of myofibroblasts was shown also by De Wever et al[14] using human colon cancer cells from established cell lines and stromal cells isolated from surgical colon cancer fragments or from normal mucosa at some distance from the tumor. Explants from such fragments yielded myofibroblasts and fibroblasts in case of tumor and normal mucosa respectively. In 48-hour cultures, the colon cancer cells invaded into the collagen only when myofibroblasts but not fibroblasts were added to the collagen. The pro-invasive activity was found also with conditioned media from myofibroblast but not from fibroblast cultures instead of the cells themselves. This activity of myofibroblasts is in line with others' experiments in vitro and in vivo; it is compatible with observations on human colon and other cancers.

Myofibroblasts are present in the stroma of many malignant tumors and they are frequently localized at the front of invasion.[4] Their presence has been positively correlated with poor prognosis.[15] That myofibroblasts may participate at the transition from the noninvasive toward the invasive phenotype is compatible with their appearance in benign lesions that have a high risk of progression toward invasive cancer. In CIN (cervical intraepithelial neoplasia), α-SMA-positive stromal cells were considered as a sign of imminent invasion.[16,17] Similarly, α-SMA-positive pericryptal fibroblasts were rare in low risk tubular adenomas but abundant in villous adenomas and in FAP (familial adenomatous polyposis) hyperplasia, both carrying high risk of malignant progression.[18]

In animal experiments, fragments of CC531 rat colon adenocarcinoma produced encapsulated noninvasive tumors when transplanted into the subcutaneous connective tissue. When the transplantation was done into experimentally induced granulation tissue, the tumors were invasive with dispersed strands of cancer cells protruding into the surrounding matrix.[19] This observation has served the development of fruitful models for the analysis of invasion in vivo.[1] Since the granulation tissue contained macrophages, other inflammatory cells and numerous capillaries next to myofibroblasts, the latter could not be held exclusively responsible for the induction of invasion. More direct evidence was obtained from the formation of invasive cancers upon subcutaneous coinjection of human colon cancer cells HCT-8 with myofibroblasts

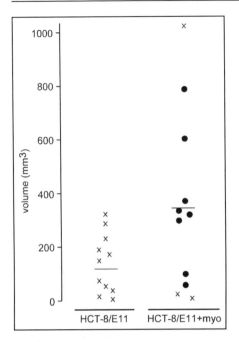

Figure 1. Growth and invasion of tumors in the flank of nude mice after subcutaneous injection of human colon cancer cells HCT-8/E11 alone or mixed with myofibroblasts (+myo). Crosses indicate noninvasive tumors, dots invasive ones. Horizontal bars indicate mean volumes, that are significantly different for both groups (p < 0.05; unpaired T-test).

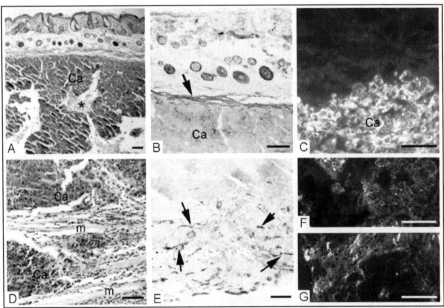

Figure 2. Histological sections from a noninvasive tumor after injection of HCT-8/E11 cells alone (panels A-C) and from invasive ones after injection of HCT-8/E11 cells plus myofibroblasts (panels D-G). Parafin sections (panels A,B,D,E) were stained with hematoxylin and eosin (panels A,D) or immunostained with an antibody against α-SMA (panels B and E). Frozen sections were immunostained with antibodies against cytokeratin (panels C,F) or against α-SMA (panel G). Panel F and G show consecutive sections from the same invasive tumor. Ca, cancer cells; *necrosis; arrows point to myofibroblasts. Scale bars = 100 mm.

into nude mice (Figs. 1, 2). The role of the murine host stromal cells in the latter experiments has not been investigated. With regard to this issue, it is interesting that the same HCT-8 cells did produce invasive cancers with lymph node metastases when transplanted alone orthotopically into the caecum and noninvasive ones in the subcutis.[20]

As observed with colon cancer, primary cultures from normal breast tissue and from carcinomas yielded respectively fibroblasts and myofibroblasts.[21] When alone inside collagen, breast cancer cells formed smooth-edged spheres but they invaded into the collagen when cocultured with myofibroblasts.[22] By contrast, fibroblasts from an embryonic lung cell line induced the epithelioid organization of T84 enterocytes inside collagen.[23] This illustrates that the source of the fibroblasts and the coculture conditions[24] are crucial for the interaction with cancer cells. It is important that, in the natural situation in vivo, the tumor is not composed solely of cancer cells and tumor-associated host fibroblast; the latter may recruit blood monocytes[25,26] which, next to their immunomodulatory activity, may also stimulate invasion.[3]

Cancer Cell-Derived TGF-β Converts Fibroblasts into Myofibroblasts

In the above mentioned experiments with human colon cancer cells,[14] differences in pro-invasive activity between normal mucosa-derived fibroblasts and colon cancer-derived myofibroblasts disappeared upon longer incubation of the collagen cocultures. Indeed, after 2 weeks of culture, the human colon cancer cells had invaded collagen gels admixed with fibroblasts as well as those admixed with myofibroblasts. In such longer cocultures, the morphotype of the fibroblasts changed and they became α-SMA-positive. From others' experiment,[21,27,28] it was suspected that a signal from the cancer cells converted the fibroblasts into myofibroblast which in their turn made the cancer cells invasive. TGF-β was the obvious candidate signal. Fibroblasts, migrating from vascular structures in vitro were converted to myofibroblasts by breast carcinoma cells MCF-7 or their conditioned medium.[21] These authors showed that TGF-β was the factor responsible for this conversion. In resting fibroblast cultures deprived of serum, α-SMA was induced without concurrent cell proliferation by addition of TGF-β or conditioned medium of MCF-7 cell cultures. The activity of the conditioned medium was neutralized by antibody against TGF-β.[27] In 3-D cultures cancer cells and TGF-β induced α-SMA but the response was more variable than in 2-D cultures on solid substrate.[29] In rats, subcutaneous injection of TGF-β resulted in the formation of granulation tissue in which α-SMA expressing myofibroblasts were particularly abundant.[21] With human colon fibroblasts,[14] the induction of α-SMA was dependent upon the time of treatment and the concentration of TGF-β; the α-SMA-inducing activity of conditioned media from colon cancer cell cultures was neutralized by an antibody against TGF-β (Fig. 3).

These results strongly suggest that fibroblasts are precursors of tumor-associated myofibroblasts; they are, however, not the only ones. Other precursors to be considered comprise preexisting myofibroblasts, CD34-positive stem cells, smooth muscle cells and pericytes. In cocultures of epithelial organoids with purified fibroblasts, vascular smooth muscle cells, pericytes or myoepithelial cells, all recently isolated from breast cancer, vascular smooth muscle cells and pericytes participated at the formation of myofibroblasts beit to a lesser extent than fibroblasts.[22] Circulating fibrocytes, positive for collagen, vimentin and CD34, derived from myeloid precursors were described by Bucala et al.[30] Such fibrocytes were analyzed in human cancers as compared to normal tissues by pathologists at the Philipps University in Marburg. CD34-positive cells were present in normal tissues and decreased in cancer whereas α-SMA-positive cells were rare in normal tissues and increased in cancer. This was found in the pancreas, breast, uterine cervix and the upper respiratory tract.[17,31-33] What causes the disappearance of the CD34-positive cells from the cancers? These cells may undergo apoptosis through a soluble factor secreted by ductal carcinoma in situ of the breast. Alternatively, CD34-positive cells may be converted to myofibroblasts through upregulation of α-SMA and downregulation of CD34 as suggested by the coexpression of both markers in subepithelial cells adjacent to high grade premalignant CINIII lesions of the uterine cervix. The participation of

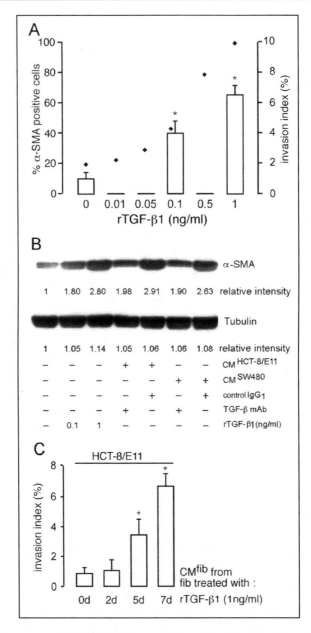

Figure 3. Colon cancer cell-derived TGF-β transdifferentiates colon fibroblasts into pro-invasive myofibroblasts. A) Concentration-dependent effect of a 7-day pretreatment with recombinant (r)TGF-β1 (abscissa) on the expression of α-SMA (diamonds and left side ordinate) and on the pro-invasive activity of the fibroblasts on HCT-8/E11 colon cancer cells (bars and right side ordinate; *significantly different from the pro-invasive activity of untreated fibroblasts in three experiments). B) Western blot of total lysates from fibroblasts treated (+) or not (-) as indicated during 7 days, immunoprobed for a-SMA followed by stripping and reprobing for γ-tubulin as a loading control. C) Time-dependent effect of pretreatment with 1 ng/ml rTGF-β1 (abscissa) on the pro-invasive activity of the fibroblast (fib) conditioned medium (CM); *significantly different from the pro-invasive activity of untreated fibroblasts in three experiments.

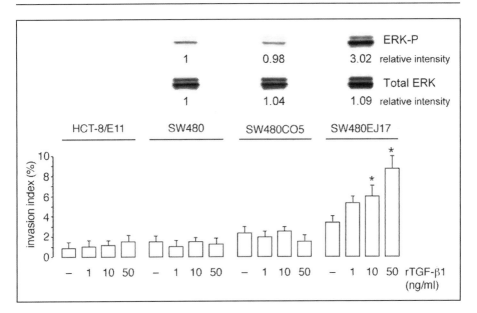

Figure 4. Effect of recombinant (r)TGF-β on the invasion of colon cancer cells into collagen as scored after 24 hours. SW480 cells were transfected with wild type Ras (CO5) or with oncogenic Ras (EJ17). Succesful oncogenic Ras transfection was assessed by activation of ERK as evidenced by dually phosphorylated p44/p42ERK in Western blots. Blots were stripped and reprobed for total ERK. *significantly different from untreated cultures.

bone marrow-derived myofibroblasts in tumorigenesis was shown experimentally in SCID mice reconstituted with bone-marrow cells and transplanted with Capan-1 pancreatic cancer cells.[34] Petersen et al[35] put forward that the cancer cells themselves may be at the origin of tumor-associated myofibroblasts. HBFL-1 cells derived from an explanted breast cancer biopsy are immortal and bear the same nonrandom X-chromosome inactivation pattern as the epithelial tumor of origin. They, however, resemble normal, finite-life-span fibroblasts by morphology, lack of tumor formation in nude mice and the ability to transit to myofibroblasts.

The above mentioned observations support the opinion that TGF-β is an indirect pro-invasive factor for epithelial cancer cells, as it converts fibroblasts that do not stimulate invasion into myofibroblasts that do stimulate invasion. We note that other factors may also be implicated in this conversion as reviewed earlier (see Table 1 in De Wever and Mareel).[4] TGF-β may also promote the invasion of cancer cells in a direct way, though this seems to be the exception rather than the rule.[36] In our experiments, the expression of oncogenic Ras made the cancer cells sensitive to the direct pro-invasive activity of TGF-β (Fig. 4) in line with others' observations.[37,38]

Myofibroblasts Are Pro-Invasive through the Combined Action of Tenascin C (Tn-C) and Hepatocyte Growth Factor/Scatter Factor (HGF/SF)

In our in vitro system of human colon cancer cells seeded on collagen gels, the pro-invasive activity of myofibroblasts has been ascribed to HGF/SF complemented with tenascin C (Tn-C), the latter marking the difference between fibroblasts and myofibroblasts. Neither of both induced invasion on its own; antibodies against either of both alone interfered with invasion. The positive invasion signaling pathway implicated inhibition of RhoA through Tn-C binding to the EGF-receptor and activation of Rac through HGF/SF binding to its c-Met receptor.[14]

HGF/SF is a heterodimeric glycoprotein composed of a 60 kDa α-chain and a 30 kDa β-chain linked by a disulfide bond. It serves as a pleiotropic cytokine that promotes survival, proliferation and morphogenesis in a wide variety of cells.[39] It was originally purified from human plasma as HGF because of its morphogenic effect on hepatocytes.[40] Later SF was identified as a motogenic agent in conditioned medium from fibroblastic cell cultures.[41] Finally, Weidner et al[42] found that HGF and SF were identical.

HGF/SF binds to and signals through a unique transmembrane tyrosine kinase receptor, the product of the c-Met proto-oncogene, responding to all criteria of an invasion and metastasis promoter.[43] In general, the c-Met protein is expressed in epithelial cells of various organs, whereas the mRNA expression and protein synthesis of HGF/SF has been detected primarily in stromal cells including fibroblasts, smooth muscle cells, and endothelial cells.[44] Upon stimulation by HGF/SF, c-Met is tyrosine-phosphorylated and this causes several signaling cascades and cellular responses.

HGF/SF meets the criteria of a pro-invasive agent in clinical and in experimental cancer. There is a vast amount of recent publications to support this statement. HGF/SF was highly expressed in invasive cancer and its higher level, evaluated by immunohistochemistry of tissue sections or by ELISA of patients' serum, served as a marker of higher probability of metastasis and worse prognosis. Recent examples were published for cancers of the colon,[45] stomach,[46] mouth mucosa[47] and salivary gland.[48] In these examples, immunohistochemistry localized HGF/SF in the stroma. In families of cell lines derived from the same normal or cancerous tissue, production of HGF/SF marked the difference between invasive and noninvasive variants in an autocrine way or between variants with and without pro-invasive activity in a paracrine way. Canine kidney epithelial MDCK cells transfected with the SV40 LT oncogene were invasive through the creation of an autocrine HGF loop.[49] Similarly, in a murine mammary epithelial cell line expressing activated M-Ras autocrine HGF was responsible for invasion-associated phenotypic changes, namely anchorage-independent growth, secretion of matrix metalloproteinases and penetration into Matrigel.[50] In malignant peripheral nerve sheath tumors (MPNST) cell variants, expression of HGF activator, converting pro-HGF into HGF, created a pro-invasive autocrine loop with constitutive activation of the c-Met receptor.[51] Examples of paracrine loops are available as well.

Myofibroblast-conditioned medium induced cell scattering and invasion of hepatocellular carcinoma cells HepG2 and HuH7 into Matrigel; these effects were mimicked by recombinant HGF, enhanced by the serine proteinase inhibitor TFPI-2 (tissue factor pathway inhibitor) and neutralized with antibody against HGF.[52,53] Human bone marrow endothelial cells secrete HGF and this significantly enhanced the invasion of multiple myeloma cells through Matrigel.[54] The combined action of HGF/SF with Tn-C described above emphasizes the need of HGF/SF for an appropriate context in order to exert its pro-invasive activity. Other strategies to make HCT-8 cells sensitive to the pro-invasiev activity of HGF/SF are overexpression of exogenous wild type STAT3[55] or activation of the canonical Wnt/APC/β-catenin signaling pathway.[56] Conversely, overexpression of sprouty-2 tempers downstream signalling of c-Met, inhibiting anchorage-independent growth, migration and invasion.[57] The response to HGF/SF signaling depended also on environmental factors such as specific extracellular matrix (ECM) proteins or hypoxia. HGF/SF stimulated MDCK cells to form tubules when embedded in collagen matrices, but not so in Matrigel.[58] Furthermore, stromal-derived epimorphin primes mammary epithelial cells for the morphogenic effect of HGF/SF in collagen type I.[59] Hypoxia inducible factors (HIFs), sensing low oxygen levels, stimulated the production of c-Met in cancer cells and, therefore, increased the invasive response to HGF/SF.[60-62]

Tn-C was originally visualized in electron micrographs as a multimeric, six-armed structure called hexabrachion.[63] It is a high-molecular-weight, multifunctional ECM glycoprotein with large disulfide-linked subunits composed of multiple structural modules. Each arm of the hexabrachion harbors a Tn assembly domain, which allows individual Tn-C polypeptides to interact at their amino termini. Tn-C contains 14.5 epidermal growth factor (EGF)-like repeats that interact with the EGF-receptor.[64] Adjacent to the EGF-like repeats are 15 fibronectin type III domains (FN-III), which constitute a highly elastic region sensitive to rapid stretching and refolding.[65] The most distal region of the Tn-C molecule contains a globular fibrinogen-homology domain which harbors a calcium-binding loop.[66] This fibrinogen domain has been found to interact with other ECM and cell surface proteins, including collagen fibrils, integrins, and heparin.[66] The different protein modules are lined up like beads on a rope and give rise to long and extended molecules. Splice variants result from alternative splicing of the FN-III repeats.[67]

In the experiments of De Wever et al,[14] Tn-C did not come as a surprise since it has been considered as a promoter of cancer malignancy since almost two decades.[68,69] Moreover, like the human colon cancer cells HCT-8, the breast cancer cells MCF-7 induced production of Tn-C by fibroblasts and this could be mimicked by TGF-β.[70] There are several lines of evidence that Tn-C may act as pro-invasive molecule, though this aspect was less investigated than for HGF/SF. Tn-C was found, as a rule, in invasive cancers but not in the normal precursor tissue. Examples are published recently for breast[71] and vulva.[72] The relationship between expression of Tn-C and prognosis is, however, less straightforward. For gastric cancer, strong expression was correlated with a significantly better 5-year survival than negative-to-moderate expression. There was no correlation with other prognostic markers, including nodal metastasis, Lauren classification and tumor size.[73] For oral and pharyngeal squamous cell carcinomas there was overexpression of Tn-C but without prognostic significance.[74] For colon cancer[75] and for melanoma,[76] lower expression or absence of Tn-C was a sign of better prognosis.

Experimentally, Tn-C affects cellular activities that are associated with the invasion program, namely cell-cell and cell-substrate adhesion, migration and anoikis. Disruption of cell-cell contact by Tn-C was observed in human mammary cancer cells T47D and MCF-7[70] and this was confirmed by later pictures showing MCF-7 cells adopting a cobble-stone-like epithelioid morphology when cultured on a fibronectin-coated substrate and scattering on Tn-C.[77] In the presence of Tn-C or of other matricellular proteins like thrombospondin 1 & 2 or SPARC (secreted protein, acidic and rich in cysteine), cells were brought into an intermediate state of adhesion to the substrate that induced survival signals to prevent anoikis and allowed cells to locomote. Signaling for de-adhesion implicated loss of actin stress fibers and focal adhesions.[78] Tn-C supported lymphocyte rolling[79] and fibroblast migration.[80] The pro-invasive signaling pathways of Tn-C necessitate further study. Enhanced migration in Boyden chambers, occurring in CHO cells transfected with Tn-C cDNA is neutralized by antibody against the FNIII domain.[67] Similarly held responsible for Tn-C-mediated loss of focal adhesion and stimulation of migration of the bovine endothelial cells GM7373, is the interaction between the TNfnA-D domain and annexinII, a cytoplasmic protein secreted to the extracellular phase and serving as a receptor. Further transduction of the signal through the plasma membrane by this extracellular protein might implicate Ca^{++} channels or an, as yet unknown, transmembrane receptor.[81] Disturbed cell-matrix adhesion was ascribed to interference of Tn-C with the binding of integrin $\alpha_5\beta_1$ to fibronectin possibly through disturbance of syndecan-4.[69,82] The use of antibody-neutralization and dominant negative transfectants pointed to the EGF-like repeats binding to the EGF-receptor for myofibroblasts-mediated invasion of human colon cancer cells.[14] Similarly, the EGF-like repeats were considered to be counteradhesive for fibroblasts, neurons and glia, and to be involved in neuronal migration and axon pathfinding during development.[83,84] By contrast, in conditioned medium from transfectants of CHO cells the EGF-like domain, unlike the FNIII domain, was not sufficient to enhance chemotactic migration of Tn-C-null mouse mammary cancer cells.[67] In favour of EGF-receptor signaling is the inactivation of the small GTPase RhoA, explaining loss of stress fibers and weakening of cell-substrate adhesion.[85]

Myofibroblasts Are Themselves Invasive

Considering the histology of invasive tumors and the activities of the composing cells it is justified to ask the question who is invading whom. Are myofibroblasts invasive? Do myofibroblasts and cancer cells respond to the same pro-invasive factors? Are myofibroblasts chemoattracted by the cancer cells? In response to signals produced by cancer cells, the tumor stroma is invaded by several cell types. The best characterized example is the hypoxia-dependent angiogenic switch in which host cell-derived endothelial cells invade into the tumor stroma to form new blood vessels.[2] Myofibroblasts also invade the tumor site and this invasion may preceed angiogenic invasion. Solid tumor implantation in transgenic mice expressing GFP under the control of the VEGF promoter leads to induction of VEGF promoter activity in myofibroblast-like cells. Subsequently, GFP-positive myofibroblast-like cells invade the whole tumor site.[86] During avascular growth of developing hepatic metastases, myofibroblast-like cells are already present, before endothelial cell recruitment.[87] De Wever et al[88] studied the invasion mechanism in collagen type I of myofibroblasts treated with colon cancer cell-derived TGF-β1. They found that N-cadherin activity was implicated in TGF-β stimulated invasion. Interestingly, TGF-β causes accumulation of N-cadherin at the tip of myofibroblast filopodia. N-cadherin belongs to the classical cadherins which are single-pass transmembrane glycoproteins that function as membrane-spanning macromolecular complexes.[89] The cadherin ectodomains mediate mainly homophilic ligation and adhesive recognition, whereas the cytoplasmic tail interacts with proteins capable of functionally linking cadherin adhesion to the actin cytoskeleton and cell-signaling pathways. Schematics of N-cadherin signaling are presented by De Wever and Mareel, and by Derycke and Bracke.[4,89]

The above mentioned experiments demonstrate that TGF-β stimulates migration on solid substrate and invasion into collagen of myofibroblasts and that N-cadherin is a necessary element in the pro-invasive loop. Since a comparison between fibroblasts and myofibroblast migration and invasion was not made, conclusions about the role of α-SMA are hard to draw. From current views on the organization of the actin cytoskeleton one would expect α-SMA to act as a contra-invasive element anchoring cells to the substrate through stress fibers at focal contacts. Retardation of motility on collagen-coated substrate was abrogated by depletion of α-SMA, for example through electroporation of monoclonal antibody against the NH2 terminal Ac-EEED sequence, causing failure of vinculin, talin and β1-integrin to organize into focal contacts.[90] Using a proteomic approach by 2-D gel electrophoresis with human primary fetal lung cell line cells HFL-1, Malmström et al[91] found a list of molecules upregulated by TGF-β. Amongst those are actin-associated proteins that are possibly implicated in invasion.[92,93]

The Pro-Invasive Switch in the Cross-Signaling Pathway

Our experiments with human colon cancer cells in an in vitro coculture system[14] revealed a pro-invasive cross-signaling between cancer cells and myofibroblasts in which the production of the matricellular protein Tn-C emerged as a switch between the invasive and the noninvasive state (Fig. 5). Even limiting ourselves to this artificial situation, we would not like to conclude that Tn-C constitutes the crucial pro-invasive alteration, given the multitude of positive and negative invasion pathways, all harbouring molecules through which such switches could be realized.[55,56,94-97] Observations on human cancers and experimental data point to myofibroblasts as an interesting starting point to understand the clinically crucial emergence of invasion. In pathological noncancer situations such as wounds, myofibroblasts emerge temporarily and undergo apoptosis when the wound has healed.[98,99] Prevention of apoptosis through the expression of anti-apoptotic genes like Bcl-2 led to the formation of hypertrophic scars composed of thick bundles of myofibroblasts.[100]

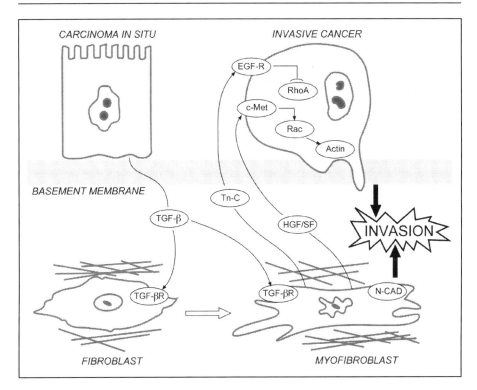

Figure 5. Schematic representation of the molecular pro-invasive cross-signaling between cancer and host cells in the microenvironment of a primary cancer. Acronyms in alphabetical order, are: EGF-R, epidermal growth factor-receptor; HGF/SF, hepatocyte growth factor/scatter factor; N-cadherin, Neural-cadherin; Rac, Ras related C3 botulinum toxin substrate; Rho, Ras homolog; Tn-C, tenascin-C; TGF-β, transforming growth factor-β; TGF-βR, transforming growth factor-β receptor.

Interestingly, myofibroblasts are sensitive targets to drugs that can be used clinically.[4,101] Despite its multifunctional character, its presence in most normal tissues and its production by various host cells, TGF-β may be a good start to analyze the molecular aspects of the pro-invasive switch since it is sensitive to modulation at multiple levels.[102] Like TGF-β, HGF activity is fine-tuned following the particular tissue contexts.[58] The conversion of pro-HGF to active HGF depends upon the balance between HGF-activator and its inhibitors HAI-1 and -2.[103] Another powerful inhibitor is NK4, a four kringle fragment of HGF, able to antagonize host cell mediated-invasion in experimental systems.[104,105] Hepatic expression of NK4 through hydrodynamics-based plasmid delivery to mice decreased invasion of murine colon cancer cells inoculated intraspenically.[106]

We conclude that, in epithelial tumors, myofibroblasts together with other host cells, participate at a pro-invasive switch, that is sensitive to modulation via various and possibly redundant molecular strategies.

Acknowledgements

G. De Bruyne and J. Roels are gratefully acknowledged for technical assistance and for preparation of the illustrations. Olivier De Wever was funded by the European Community project METABRE (contract LSHC-CT-2004-506049).

References

1. Mueller MM, Fusenig NE. Friends or foes - bipolar effects of the tumour stroma in cancer. Nat Rev Cancer 2004; 4:839-849.
2. Carmeliet P, Jain RK. Angiogenesis in cancer and other diseases. Nature 2000; 407:249-257.
3. Opdenakker G, Van Damme J. The countercurrent principle in invasion and metastasis of cancer cells. Recent insights on the roles of chemokines. Int J Dev Biol 2004; 48:519-527.
4. De Wever O, Mareel M. Role of tissue stroma in cancer cell invasion. J Pathol 2003; 200:429-447.
5. Paget S. The distribution of secondary growths in cancer of the breast. Lancet 1889; 1:571-573.
6. Allinen M, Beroukhim R, Cai L et al. Molecular characterization of the tumor microenvironment in breast cancer. Cancer Cell 2004; 6:17-32.
7. Howe JR, Roth S, Ringold JC et al. Mutations in the SMAD4/DPC4 gene in juvenile polyposis. Science 1998; 280:1086-1088.
8. Bardeesy N, Sinha M, Hezel AF. Loss of the Lkb1 tumour suppressor provokes intestinal polyposis but resistance to transformation. Nature 2002; 419:162-167.
9. Young J, Biden KG, Simms LA et al. HPP1: a transmembrane protein-encoding gene commonly methylated in colorectal polyps and cancers. Proc Natl Acad Sci USA 2001; 98:265-270.
10. Kinzler KW, Vogelstein B. Landscaping the cancer terrain. Science 1998; 280:1036-1037.
11. Majno G, Gabbiani G, Hirschel BJ et al. Contraction of granulation tissue in vitro: similarity to smooth muscle. Science 1971; 173:548-550.
12. Desmoulière A, Guyot C, Gabbiani G. The stroma reaction myofibroblast: a key player in the control of tumor cell behavior. Int J Dev Biol 2004; 48:509-517.
13. Dimanche-Boitrel MT, Vakaet Jr L, Pujuguet P et al. In vivo and in vitro invasiveness of a rat colon cancer cell line maintaining E-cadherin expression. An enhancing role of tumor-associated myofibroblasts. Int J Cancer 1994; 56:512-521.
14. De Wever O, Nguyen Q-D, Van Hoorde L et al. Tenascin-C and SF/HGF produced by myofibroblasts in vitro provide convergent pro-invasive signals to human colon cancer cells through RhoA and Rac. FASEB J 2004; 18:1016-1018.
15. Seemayer TA, Lagace R, Schurch W et al. Myofibroblasts in the stroma of invasive and metastatic carcinoma: a possible host response to neoplasia. Am J Surg Pathol 1979; 3:525-533.
16. Cintorino M, Bellizzi de Marco E, Leoncini P et al. Expression of a-smooth-muscle actin in stromal cells of the uterine cervix during epithelial neoplastic changes. Int J Cancer 1991; 47:843-846.
17. Barth PJ, Ebrahimsade S, Ramaswamy A et al. CD34+ fibrocytes in invasive ductal carcinoma, ductal carcinoma in situ, and benign breast lesions. Virchows Arch 2002; 440:298-303.
18. Sappino A-P, Dietrich P-Y, Skalli O et al. Colonic pericryptal fibroblasts. Differentiation pattern in embryogenesis and phenotypic modulation in epithelial proliferative lesions. Virchows Arch A Pathol Anat 1989; 415:551-557.
19. Dingemans KP, Zeeman-Boeschoten IM, Keep RF et al. Transplantation of colon carcinoma into granulation tissue induces an invasive morphotype. Int J Cancer 1993; 54:1010-1016.
20. Van Hoorde L, Pocard M, Maryns I et al. Induction of invasion in vivo of a-catenin-positive HCT-8 human colon-cancer cells. Int J Cancer 2000; 88:751-758.
21. Rønnov-Jessen L, Van Deurs B, Nielsen M et al. Identification, paracrine generation, and possible function of human breast carcinoma myofibroblasts in culture. In Vitro Cell Dev Biol 1992; 28A:273-283.
22. Rønnov-Jessen L, Petersen OW, Koteliansky VE et al. The origin of the myofibroblasts in breast cancer. Recapitulation of tumor environment in culture unravels diversity and implicates converted fibroblasts and recruited smooth muscle cells. J Clin Invest 1995; 95:859-873.
23. Halttunen T, Marttinen A, Rantala I et al. Fibroblasts and transforming growth factor-ß induce organization and differentiation of T84 human epithelial cells. Gastroenterology 1996; 111:1252-1262.
24. Kunz-Schughart LA, Heyder P, Schroeder J et al. A heterologous 3-D coculture model of breast tumor cells and fibroblasts to study tumor-associated fibroblast differentiation. Exp Cell Res 2001; 266:74-86.
25. Silzle T, Kreutz M, Dobler MA et al. Tumor-associated fibroblasts recruit blood monocytes into tumor tissue. Eur J Immunol 2003; 33:1311-1320.
26. Silzle T, Randolph GJ, Kreutz M et al. The fibroblast: sentinel cell and local immune modulator in tumor tissue. Int J Cancer 2004; 108:173-180.
27. Rønnov-Jessen L, Petersen OW. Induction of a-smooth muscle actin by transforming growth factor-ß1 in quiescent human breast gland fibroblasts. Implications for myofibroblast generation in breast neoplasia. Lab Invest 1993; 68:696-707.

28. Desmoulière A, Geinoz A, Gabbiani F et al. Transforming growth factor-ß1 induces a-smooth muscle actin expression in granulation tissue myofibroblasts and in quiescent and growing cultured fibroblasts. J Cell Biol 1993; 122:103-111.

29. Kunz-Schughart LA, Wenninger S, Neumeier T et al. Three-dimensional tissue structure affects sensitivity of fibroblasts to TGF-β1. Am J Physiol Cell Physiol 2003; 284:C209-C219.

30. Bucala R, Spiegel LA, Chesney J et al. Circulating fibrocytes define a new leukocyte subpopulation that mediates tissue repair. Mol Med 1994; 1:71-81.

31. Barth PJ, Ebrahimsade S, Hellinger A et al. CD34+ fibrocytes in neoplastic and inflammatory pancreatic lesions. Virchows Arch 2002; 440:128-133.

32. Ramaswamy A, Moll R, Barth PJ. CD34+ fibrocytes in tubular carcinomas and radial scars of the breast. Virchows Arch 2003; 443:536-540.

33. Barth PJ, Schenck zu Schweinsberg T, Ramaswamy A et al. CD34+ fibrocytes, a-smooth muscle antigen-positive myofibroblasts, and CD117 expression in the stroma of invasive squamous cellcarcinomas of the oral cavity, pharynx, and larynx. Virchows Arch 2004; 444:231-234.

34. Ishii G, Sangai T, Oda T et al. Bone-marrow-derived myofibroblasts contribute to the cancer-induced stromal reaction. Biochem Biophys Res Commun 2003; 309:232-2435.

35. Petersen OW, Nielsen HL, Gudjonsson T et al. Epithelial to mesenchymal transition in human breast cancer can provide a nonmalignant stroma. Am J Pathol 2003; 162:391-402.

36. Brown KA, Aakre ME, Gorska AE et al. Induction by transforming growth factor-ß1 of epithelial to mesenchymal transition is a rare event in vitro. Breast Cancer Res 2004; 6:R215-R231.

37. Oft M, Peli J, Rudaz C et al. TGF-β1 and Ha-Ras collaborate in modulating the phenotypic plasticity and invasiveness of epithelial tumor cells. Genes Dev 1996; 10:2462-2477.

38. Oft M, Akhurst RJ, Balmain A. Metastasis is driven by sequential elevation of H-ras and Smad2 levels. Nat Cell Biol 2002; 4:487-494.

39. Trusolino L, Comoglio PM. Scatter-factor and semaphorin receptors: cell signalling for invasive growth. Nat Rev Cancer 2002; 2:289-300.

40. Gohda E, Tsubouchi H, Nakayama H et al. Purification and partial characterization of hepatocyte growth factor from plasma of a patient with fulminant hepatic failure. J Clin Invest 1988; 81:414-419.

41. Gherardi E, Gray J, Stoker M et al. Purification of scatter factor, a fibroblast-derived basic protein that modulates epithelial interactions and movement. Proc Natl Acad Sci USA 1989; 86:5844-5848.

42. Weidner KM, Arakaki N, Hartmann G et al. Evidence for the identity of human scatter factor and human hepatocyte growth factor. Proc Natl Acad Sci USA 1991; 88:7001-7005.

43. Oliveira MJ, Mareel M, Leroy A. Cancer invasion and metastasis: cellular, molecular and clinical aspects. Encyclopedic Reference of Genomics and Proteomics in Molecular Medicine 2004 (In Press).

44. Sonnenberg E, Meyer D, Weidner KM et al. Scatter factor/hepatocyte growth factor and its receptor, the c-met tyrosine kinase, can mediate a signal exchange between mesenchyme and epithelia during mouse development. J Cell Biol 1993; 123:223-235.

45. Fukuura T, Miki C, Inoue T et al. Serum hepatocyte growth factor as an index of disease status of patients with colorectal carcinoma. Br J Cancer 1998; 78:454-459.

46. Tanaka K, Miki C, Wakuda R et al. Circulating level of hepatocyte growth factor as a useful tumor marker in patients with early-stage gastric carcinoma. Scand J Gastroenterol 2004; 39:754-760.

47. Chen Y-S, Wang J-T, Chang Y-F et al. Expression of hepatocyte growth factor and c-met protein is significantly associated with the progression of oral squamous cell carcinoma in Taiwan. J Oral Pathol Med 2004; 33:209-217.

48. Tsukinoki K, Yasuda M, Mori Y et al. Hepatocyte growth factor and c-Met immunoreactivity are associated with metastasis in high grade salivary gland carcinoma. Oncol Rep 2004; 12:1017-1021.

49. Martel C, Harper F, Cereghini S et al. Inactivation of retinoblastoma family proteins by SV40 T antigen results in creation of a hepatocyte growth factor/scatter factor autocrine loop associated with an epithelial-fibroblastoid conversion and invasiveness. Cell Growth Differ 1997; 8:165-178.

50. Zhang KX, Ward KR, Schrader JW. Multiple aspects of the phenotype of mammary epithelial cells transformed by expression of activated M-Ras depend on an autocrine mechanism mediated by hepatocyte growth factor/scatter factor. Mol Cancer Res 2004; 2:242-255.

51. Su W, Gutmann DH, Perry A et al. CD44-independent hepatocyte growth factor/c-Met autocrine loop promotes malignant peripheral nerve sheath tumor cell invasion in vitro. Glia 2004; 45:297-306.

52. Neaud V, Faouzi S, Guirouilh J et al. Human hepatic myofibroblasts increase invasiveness of hepatocellular carcinoma cells: evidence for a role of hepatocyte growth factor. Hepatology 1997; 26:1458-1466.

53. Neaud V, Hisaka T, Monvoisin A et al. Paradoxical pro-invasive effect of the serine proteinase inhibitor tissue factor pathway inhibitor-2 on human hepatocellular carcinoma cells. J Biol Chem 2000; 275:35565-35569.
54. Vande Broek I, Asosingh K, Allegaert V et al. Bone marrow endothelial cells increase the invasiveness of human multiple myeloma cells through upregulation of MMP-9: evidence for a role of hepatocyte growth factor. Leukemia 2004; 18:976-982.
55. Rivat C, De Wever O, Bruyneel E et al. Disruption of STAT3 signaling leads to tumor cell invasion through alterations of homotypic cell-cell adhesion complexes. Oncogene 2004; 23:3317-3327.
56. Le Floch N, Rivat C, De Wever O et al. The proinvasive activity of Wnt-2 is mediated through a noncanonical Wnt pathway coupled to GSK-3β and c-Jun/AP-1 signaling. FASEB J 2005; 19:144-146.
57. Lee CC, Putnam AJ, Miranti CK et al. Overexpression of sprouty 2 inhibits HGF/SF-mediated cell growth, invasion, migration, and cytokinesis. Oncogene 2004; 23:5193-5202.
58. Rosário M, Birchmeier W. How to make tubes: signaling by the Met receptor tyrosine kinase. Trends Cell Biol 2003; 13:328-335.
59. Hirai Y, Lochter A, Galosy S et al. Epimorphin functions as a key morphoregulator for mammary epithelial cells. J Cell Biol 1998; 140:159-169.
60. Bottaro DP, Liotta LA. Cancer: Out of air is not out of action. Nature 2003; 423:593-595.
61. Steeg PS. Metastasis suppressors alter the signal transduction of cancer cells. Nat Rev Cancer 2003; 3:55-63.
62. Pennacchietti S, Michieli P, Galluzzo M et al. Hypoxia promotes invasive growth by transcriptional activation of the met protooncogene. Cancer Cell 2003; 3:347-361.
63. Erickson HP, Inglesias JL. A six-armed oligomer isolated from cell surface fibronectin preparations. Nature 1984; 311:267-269.
64. Swindle CS, Tran KT, Johnson TD et al. Epidermal growth factor (EGF)-like repeats of human tenascin-C as ligands for EGF receptor. J Cell Biol 2001; 154:459-468.
65. Oberhauser AF, Marszalek PE, Erickson HP et al. The molecular elasticity of the extracellular matrix protein tenascin. Nature 1998; 393:181-185.
66. Jones FS, Jones PL. The tenascin family of ECM glycoproteins: structure, function, and regulation during embryonic development and tissue remodeling. Dev Dyn 2000; 218:235-259.
67. Tsunoda T, Inada H, Kalembeyi I et al. Involvement of large tenascin-C splice variants in breast cancer progression. Am J Pathol 2003; 162:1857-1867.
68. Chiquet-Ehrismann R, Mackie EJ, Pearson CA et al. Tenascin: an extracellular matrix protein involved in tissue interactions during fetal development and oncogenesis. Cell 1986; 47:131-139.
69. Chiquet-Ehrismann R, Chiquet M. Tenascins: regulation and putative functions during pathological stress. J Pathol 2003; 200:488-499.
70. Chiquet-Ehrismann R, Kalla P, Pearson CA. Participation of tenascin and transforming growth factor-ß in reciprocal epithelial-mesenchymal interactions of MCF7 cells and fibroblasts. Cancer Res 1989; 49:4322-4325.
71. Goepel C, Buchmann J, Schultka R et al. Tenascin - A marker for the malignant potential of preinvasive breast cancers. Gynecol Oncol 2000; 79:372-378.
72. Goepel C, Stoerer S, Koelbl H. Tenascin in preinvasive lesions of the vulva and vulvar cancer. Anticancer Res 2003; 23:4587-4591.
73. Wiksten JP, Lundin J, Nordling S et al. Tenascin-C expression correlates with prognosis in gastric cancer. Oncology 2003; 64:245-250.
74. Atula T, Hedstrom J, Finne P et al. Tenascin-C expression and its prognostic significance in oral and pharyngeal squamous cell carcinoma. Anticancer Res 2003; 23:3051-3056.
75. Sis B, Sagol O, Kupelioglu A et al. Prognostic significance of matrix metalloproteinase-2, cathepsin D, and tenascin-C expression in colorectal carcinoma. Pathol Res Pract 2004; 200:379-387.
76. Ilmonen S, Jahkola T, Turunen JP et al. Tenascin-C in primary malignant melanoma of the skin. Histopathology 2004; 45:405-411.
77. Martin D, Brown-Luedi M, Chiquet-Ehrismann R. Tenascin-C signaling through induction of 14-3-3 tau. J Cell Biol 2003; 160:171-175.
78. Murphy-Ullrich JE. The de-adhesive activity of matricellular proteins: is intermediate cell adhesion an adaptive state? J Clin Invest 2001; 107:785-790.
79. Clark RA, Erickson HP, Springer TA. Tenascin supports lymphocyte rolling. J Cell Biol 1997; 137:755-765.
80. McKean DM, Sisbarro L, Ilic D et al. FAK induces expression of Prx1 to promote tenascin-C-dependent fibroblast migration. J Cell Biol 2003; 161:393-402.

81. Chung CY, Murphy-Ullrich JE, Erickson HP. Mitogenesis, cell migration, and loss of focal adhesions induced by tenascin-C interacting with its cell surface receptor, annexin II. Mol Biol Cell 1996; 7:883-892.

82. Saoncella S, Echtermeyer F, Denhez F et al. Syndecan-4 signals cooperatively with integrins in a Rho-dependent manner in the assembly of focal adhesions and actin stress fibers. Proc Natl Acad Sci USA 1999; 96:2805-2810.

83. Fischer D, Brown-Lüdi M, Schulthess T et al. Concerted action of tenascin-C domains in cell adhesion, anti-adhesion and promotion of neurite outgrowth. J Cell Sci 1997; 110:1513-1522.

84. Götz B, Scholze A, Clement A et al. Tenascin-C contains distinct adhesive, anti-adhesive, and neurite outgrowth promoting sites for neurons. J Cell Biol 1996; 132:681-699.

85. Wenk MB, Midwood KS, Schwarzbauer JE. Tenascin-C suppresses Rho activation. J Cell Biol 2000; 150:913-919.

86. Fukumura D, Xavier R, Sugiura T et al. Tumor induction of VEGF promoter activity in stromal cells. Cell 1998; 94:715-725.

87. Olaso E, Salado C, Egilegor E et al. Proangiogenic role of tumor-activated hepatic stellate cells in experimental melanoma metastasis. Hepatology 2003; 37:674-685.

88. De Wever O, Westbroek W, Verloes A et al. Critical role of N-cadherin in myofibroblast invasion and migration in vitro stimulated by colon-cancer-cell-derived TGF-β or wounding. J Cell Sci 2004; 117:4691-4703.

89. Derycke LDM, Bracke ME. N-cadherin in the spotlight of cell-cell adhesion, differentiation, embryogenesis, invasion and signalling. Int J Dev Biol 2004; 48:463-476.

90. Rønnov-Jessen L, Petersen OW. A function for filamentous a-smooth muscle actin: retardation of motility in fibroblasts. J Cell Biol 1996; 134:67-80.

91. Malmstrom J, Lindberg H, Lindberg C et al. Transforming growth factor-β1 specifically induce proteins involved in the myofibroblast contractile apparatus. Mol Cell Proteomics 2004; 3:466-477.

92. De Corte V, Bruyneel E, Boucherie C et al. Gelsolin-induced epithelial cell invasion is dependent on Ras-Rac signaling. EMBO J 2002; 21:6781-6790.

93. De Corte V, Van Impe K, Bruyneel E et al. Increased importin-ß-induced nuclear import of the actin modulating protein CapG promotes cell invasion. J Cell Sci 2004; 117:5283-5292.

94. Mareel M, Leroy A. Clinical, cellular, and molecular aspects of cancer invasion. Physiol Rev 2003; 83:337-376.

95. Rivat C, Le Floch N, Sabbah M et al. Synergistic cooperation between AP-1 and LEF-1 transcription factors in the activation of the matrilysin promoter by the src oncogene: implications in cellular invasion. FASEB J 2003; 17:1721-1723.

96. Rodrigues S, Attoub S, Nguyen Q-D et al. Selective abrogation of the proinvasive activity of the trefoil peptides pS2 and spasmolytic polypeptide by disruption of the EGF receptor signaling pathways in kidney and colonic cancer cells. Oncogene 2003; 22:4488-4497.

97. Rodrigues S, Van Aken E, Van Bocxlaer S et al. Trefoil peptides as proangiogenic factors in vivo and in vitro: implication of cyclooxygenase-2 and EGF receptor signaling. FASEB J 2003; 17:7-16.

98. Dvorak HF. Tumors: wounds that do not heal. N Engl J Med 1986; 315:1650-1659.

99. Desmoulière A, Redard M, Darby I et al. Apoptosis mediates the decrease in cellularity during the transition between granulation tissue and scar. Am J Pathol 1995; 146:56-66.

100. Teofoli P, Barduagni S, Ribuffo M et al. Expression of Bcl-2, p53, c-jun and c-fos protooncogenes in keloids and hypertrophic scars. J Dermatol Sci 1999; 22:31-37.

101. De Wever O, Mareel M. Role of myofibroblasts at the invasion front. Biol Chem 2002; 383:55-67.

102. Derynck R, Akhurst RJ, Balmain A. TGF-β signaling in tumor suppression and cancer progression. Nat Genet 2001; 29:117-129.

103. Parr C, Watkins G, Mansel RE et al. The hepatocyte growth factor regulatory factors in human breast cancer. Clin Cancer Res 2004; 10:202-211.

104. Tanaka T, Shimura H, Sasaki T et al. Gallbladder cancer treatment using adenovirus expressing the HGF/NK4 gene in a peritoneal implantation model. Cancer Gene Ther 2004; 11:431-440.

105. Ohuchida K, Mizumoto K, Murakami M et al. Radiation to stromal fibroblasts increases invasiveness of pancreatic cancer cells through tumor-stromal interactions. Cancer Res 2004; 64:3215-3222.

106. Wen J, Matsumoto K, Taniura N et al. Hepatic gene expression of NK4, an HGF-antagonist/angiogenesis inhibitor, suppresses liver metastasis and invasive growth of colon cancer in mice. Cancer Gene Ther 2004; 11:419-430.

CHAPTER 9

Proangiogenic Implications of Hepatic Stellate Cell Transdifferentiation into Myofibroblasts Induced by Tumor Microenvironment

Elvira Olaso, Beatriz Arteta, Clarisa Salado, Eider Eguilegor, Natalia Gallot, Aritz Lopategi, Virginia Gutierrez, Miren Solaun, Lorea Mendoza and Fernando Vidal-Vanaclocha*

Abstract

Hepatic stellate cells are perisinusoidal fibroblasts that transdifferentiate into myofibroblasts in response to paracrine factors released from cancer cells and cancer-activated endothelial cells. Tumor-associated myofibroblasts exhibit contractility, proliferation, production of extracellular matrix molecules and metalloproteases. They secrete soluble factors inducing proinflammatory and immune suppressant effects. Myofibroblasts are present in avascular micrometastasis prior to endothelial cell recruitment, and act as supporting stroma for tumor neoangiogenesis. In replacement-type cancer growth, the reticular arrangement of tumor-associated myofibroblasts provides a sinusoidal-type angiogenic pattern. In pushing-type cancer growth, fibrous tract–forming myofibroblasts support a portal-type angiogenic pattern. Additionally, tumor-activated myofibroblasts support cancer development via paracrine release of tumor invasion and proliferation-stimulating factors. In summary, this information suggests that the ability of cancer cells to activate hepatic stellate cells and to collaborate with myofibroblasts along the metastatic process may represent a key phenotypic property of liver-colonizing cancer cells. On the other hand, experimental anti-tumor and anti-angiogenic agents have inhibited intrametastatic recruitment and proangiogenic activities of hepatic myofibroblasts, suggesting that targeting tumorigenic effects of these cells may contribute to hepatic metastasis inhibition.

Introduction

Genetic mutations within cancer cells results in deregulation of their normal growth-controlling mechanisms. Thus, tumorigenesis has long been considered as a cell-autonomous process governed by the genes carried by cancer cells. However, recent research has revealed that

*Corresponding Author: Fernando Vidal-Vanaclocha—Basque Country University School of Medicine and Dentistry, Department of Cell Biology, Leioa, Bizkaia 48940, Spain; and Dominion Pharmakine Ltd., Bizkaia Technology Park, Bldg 801, First Floor, Derio, Bizkaia 48160, Spain. Email: gcpvivaf@lg.ehu.es

Tissue Repair, Contraction and the Myofibroblast,
edited by Christine Chaponnier, Alexis Desmoulière and Giulio Gabbiani.
©2006 Landes Bioscience and Springer Science+Business Media.

cancer cell growth is just part of the story in cancer development. Mounting evidence now suggests that a cancer cell interacts with its local and systemic microenvironments, and each of them profoundly influences the behavior of the other. These tumor-host interactions permit, and even encourage, cancer progression as reviewed by Cheng and Weiner.[1] For example, the microenvironment created by the connective tissue stroma in which cancer cells develop affects tumor growth and invasiveness, and even the ability of cancer cells to metastasize as reported by Picard et al,[2] Chung et al[3] and Grégoire et al.[4] Moreover, paracrine signaling between cancer cells and neighboring stromal fibroblasts also contributes to proangiogenic activities developed in the cancer microenvironment as reviewed by Elenbaas and Weinberg.[5] In the liver, primary and secondary malignant tumors are frequently infiltrated by myofibroblasts, as noted by Schürch et al[6] and Cornil et al.[7] However, their origin and functional implications remain unclear. In this review, we summarize our recent results on the bidirectional signaling interactions between cancer cells and neighboring stromal fibroblasts occurring during the hepatic colonization of experimental tumors. On this basis, we propose a model to understand the contribution of hepatic tumor microenvironment to the transdifferentiation of perisinusoidal hepatic stellate cells (HSCs) into myofibroblasts endowed with proangiogenic capabilities. Finally, we discuss the therapeutic potential of targeting the transdifferentiation process and tumorigenic activities of resulting tumor-activated hepatic myofibroblasts.

Cancer Microenvironment and Tumor-Activated Myofibroblasts

Cancer progression is only possible through the close interaction of neoplastic and non-neoplastic cells. The cyto-architecture of a malignant solid tumor consists of a mixture of cancer cells and host cells in variable proportions. They form a poorly organized entity, the function of which is maintained by the dynamic interplay between cancer and host cells. A cooperative relationship among these cells occurs, with each cell type deriving functional advantages from the others. Tumor-associated host cells form a connective tissue stroma enriched with microvessels and infiltrating inflammatory cells. Altogether, they form the tumor stroma, which is conceptually restricted to those non-neoplastic cells providing physical and functional support to cancer cells, including the intratumoral trafficking of molecules and cells. In contrast, 'cancer microenvironment' implies the functional and structural constellation of both neoplastic and non-neoplastic cells and their extracellular components, with emphasis on their functional interactions. Therefore, cancer microenvironment not only includes the structural components of the tumor stroma, but also cytokines, chemokines, and growth factors that may derive from either cancer or host cells. Recently, the functional effects caused by these molecules have made the cancer microenvironment an important focus for research, both in the laboratory and in the clinic (for more information, see http://cancermicroenvironment.tau.ac.il).[8]

Myofibroblasts are usually the predominant cell type in the stroma of both primary and metastatic tumors. They are phenotypically different from normal fibroblasts. They are capable of inducing tissue contraction due to their expression of α-smooth muscle actin. They exhibit a high proliferation rate and secrete large amounts of extracellular matrix molecules. In fact, they are largely responsible for the desmoplasia observed in carcinomas, as observed by Musso et al,[9] and Theret et al.[10] They also produce factors promoting the proliferation of the nearby cancer cells and even regional inflammatory response and immune suppression as reviewed by Silzle et al.[11] Recently, Nakagawa et al[12] have reported distinct molecular expression profiles of cancer-associated fibroblasts in colon cancer metastasis in the liver which support the notion that fibroblasts form a favorable microenvironment for cancer cells. Altogether, these properties account for the enhanced tumor-promoting activity of myofibroblasts isolated from tumor tissue, in comparison to the fibroblasts obtained from normal tissues.

Similar to primary neoplasm, metastasis is also an aberrant tissue-reconstitution process involving cancer cells and several host cell types. Myofibroblasts are also present in metastases wherever they develop. Again, metastatic cells depend on stromal cells from the specific microenvironment of target organ for their optimal growth. Moreover, in agreement with the

Seed and Soil Theory, postulated by Stephan Paget[13] in 1889, it is likely that the organ-specific nature of the stromal microenvironment contributes to which organ will provide a hospitable environment for the seeding and eventual outgrowth of metastatic cells (for a review, see Fidler[14]).

Pathophysiologic Aspects of the Hepatic Metastasis Process

Due to the efficient hepatic filtration of venous drainage from abdominal viscera and of a cardiac output fraction, the retention of circulating cancer cells within the hepatic microvasculature is an early event for most of malignant tumors. However, development of clinically relevant metastases is the final outcome of a complex balance between pro-metastatic and anti-metastatic actions occurring at cancer cell implantation sites in the liver. This process begins by cancer cell trapping at terminal portal venules and proximal segments of sinusoids, as reported by Vidal-Vanaclocha et al,[15] leading to the destruction of a significant percentage of cancer cells by mechanical stress. Other indirect microvascular events further contribute to the proinflammatory response to cancer cells, such as reoxygenation of ischemic sinusoidal segments subsequent to transient micro-infarcts in the periportal sinusoids due to the cancer cells trapping process—see Jessup et al[16]—and sinusoidal cells damage by the influx and destruction of cancer cells (see Weiss).[17] Next, a reciprocal functional interaction occurs between surviving cancer cells and hepatic sinusoidal cells leading to an acute inflammatory response of the latter cells, as described by Vidal-Vanaclocha et al.[18] An additional killing of cancer cells occurs in the following hours, as a result of the release of nitric oxide, reactive oxygen intermediates, and some specific lymphotoxins by LSECs, Kupffer cells and liver-associated lymphocytes, as reported by Wang et al,[19] and Vekemans et al[20] and Luo et al.[21] Despite these innate defense resources of the hepatic microvasculature against tumor growth, some cancer cells neutralize, overcome or resist to this antitumor microenvironment. For example, it has been reported that local release of immune suppressant factors such as soluble ICAM-1, PGE-2, IL-10 and Fas ligand, appear to promote regional immune-tolerance and cancer cell immune-escape mechanisms, as suggested also by Jessup et al[22] and Cho.[23]

On this basis, some cancer cells progress along the hepatic metastasis process and form avascular micrometastases. These micrometastases appear to be sublobular compartment–specific, because they preferentially locate in periportal areas of hepatic tissue, as described by Vidal-Vanaclocha.[15] Then, a rich proangiogenic and tumor growth-stimulating stroma, generated from myofibroblasts and other tumor-activated hepatic sinusoidal cells, promote the transition from early avascular micrometastatic growth (subclinical occult metastases) to large-size vascularized metastasis (clinically-relevant), as shown by Olaso et al.[24,25] In view of these observations, a hypothesis has been advanced, where the liver is envisioned as an organ containing microenvironmental factors providing a favorable "soil" for "seeding" and growth of certain cancer cells. A more detailed knowledge on such microenvironmental factors may provide new molecular targets for therapeutic innovation.

Hepatic Stellate Cell Transdifferentiation into Myofibroblasts during the Microvascular Stage of the Hepatic Metastasis Process

Intrahepatic invasion and growth of surviving cancer cells lead to an extensive liver tissue damage that involves recruitment and activation of neighboring myofibroblast-like cells, according to Olaso et al.[24] These cells have also been identified in primary malignant tumors from the liver and are responsible for the remodeling and deposition of tumor-associated extracellular matrix. They have also been involved in the chemotactic migration and growth of both hepatoma and metastatic cancer cells, as shown by Faouzi.[26] Histological evaluation of stromal cells in murine hepatic tissue on days 3 to 6 following the intrasplenic injection of several cancer cell lines reveals the presence of α-smooth muscle actin-expressing cells around and within cancer cell colonies located at the sinusoidal area. Interestingly, no α-smooth muscle actin-expressing cells occurred in sinusoids from unaffected hepatic tissue, as reported by Olaso et al.[24] This suggests that contact-mediated and/or short-distance paracrine mechanisms are promoting the transdifferentiation of some cells into myofibroblast-like cells at sites of metastasis implantation.

The most probable cellular origin of transdifferentiated myofibroblast-like cells is the hepatic stellate cell (formerly called Ito cell or fat-storing cell). This is an organ-specific pericyte-like mesenchymal cell, located at the perisinusoidal space of Disse that accounts for 15% of cells in normal liver as reviewed by Wake.[27] Stellate cells develop long cellular processes and contain perinuclear vitamin A droplets (Fig. 1), and respond to chronic tissue injury by transdifferentiating into a myofibroblast-like cell, devoid of vitamin A droplets and expressing α-smooth muscle actin. Stellate cells are responsible for matrix production during liver fibrosis and are a primary focus of current efforts for the development of antifibrotic therapies as reviewed by Olaso and Friedman.[28] However, other sources of tumor-associated myofibroblasts such as circulating bone-marrow mesenchymal progenitors and liver epithelial-mesenchymal transitions cannot be discarded, as reported by Direkze et al[29] and Kalluri et al.[30]

According to Olaso et al,[24,25] Faouzi et al,[26] Shimizu et al,[31] Musso et al[9] and Le Pavic et al,[32] soluble factors from primary and metastatic cancer cells activate cultured hepatic stellate cells, endowing them with some of the myofibroblast-like phenotypic properties observed in stellate cells of fibrotic liver, as summarized by Olaso and Friedman.[28] They exhibit de novo α-smooth muscle actin and β-type platelet-derived growth factor receptor expression, increased

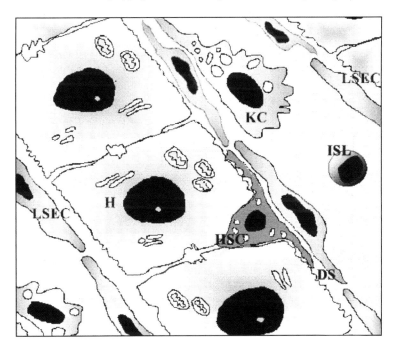

Figure 1. Cellular composition of the hepatic sinusoid microenvironment. Wall-forming sinusoidal endothelial cells (LSECs) constitute a highly-fenestrated microvascular barrier regulating transhepatic passage of circulating cells and molecules, and blood-liver exchanges. Intrasinusoidal Kupffer cell (KC) is an organ-specific resident macrophage contributing to liver immune response and scavenger functions. Perisinusoidal hepatic stellate cells (HSCs)—Ito cells or fat-storing cells—are vitamin-A-storing pericytes that produce and regulate extracelullar matrix at Disse's space (DS) and that can contribute to sinusoidal hemodynamics. Intrasinusoidal leukocytes (ISL)—i.e., natural killer cells, dendritic cells and some granulocytes—are transiently resident cells contributing to liver immune response.

cell chemotactic migration and proliferation, and enhanced matrix metalloprotease-2, tissue-inhibitor matrix metalloprotease-2, ADAM12, glycosaminoglycans and collagen type I synthesis in as compared to untreated cells. Tumor-activation of hepatic stellate cells also results in enhanced production of vascular endothelial growth factor (VEGF). In turn, this VEGF induces a proangiogenic phenotype in liver sinusoidal endothelial cells (LSECs) by promoting their chemotactic migration and survival as reported by Olaso et al.[25] Uncovered phenotypic properties of tumor-activated stellate cells have been summarized in Table 1.

Mechanistically, circulating cancer cell infiltration into the hepatic microvasculature generates a proinflammatory activation of the sinusoidal cell milieu that initially affects to endothelial cells and Kupffer cells. Tumor-derived factors such as carcinoembryonic antigen, VEGF and other proinflammatory cytokines, but also reactive oxygen intermediates such as hydrogen peroxide, appear to be involved, as reported by Thomas et al,[36] Vidal-Vanaclocha et al[18] and Mendoza et al.[37,38] In turn, paracrine factors released from tumor-activated sinusoidal cells act in cascade by targeting again cancer cells and hepatic stellate cells, (Vidal-Vanaclocha et al, unpublished data). The high permeability of fenestrated LSECs, first reported by Wisse et al,[39] may additionally facilitate a bidirectional transendothelial passage of tumor-derived soluble factors promoting hepatic stellate cell involvement in the process.

Table 1. *In situ and in vitro phenotypic properties of hepatic myofibroblasts resulting from the hepatic stellate cell transdifferentiation induced by tumor-derived factors*

Phenotypic Properties of Tumor-Activated Hepatic Stellate Cells	References
Increased proliferation rate	Olaso et al,[24,25] Faouzi etal,[26] Shimizu et al[31]
Increased chemotactic migration	Olaso et al,[24,25] Faouzi etal,[26] Shimizu et al[31]
Cytoskeletal changes (Cell spreading, generation of cytoplasmic processes, increased SMA expression)	Faouzi et al,[26] Shimizu et al[31]
Secretion of soluble tumor cell growth- and migration-stimulating factors	Olaso et al,[24,25] Shimizu et al[31]
Increased secretion of proangiogenic factors inducing endothelial cell survival, proliferation and migration	Olaso et al,[24]
Increased extracellular matrix synthesis and remodeling	Musso et al,[9] Theret et al,[10] Olaso et al,[24] Le Pavic et al,[23] Ooi et al,[33] Shimizu et al[31]
Surface receptors de novo expression (increased expression of β-PDGF, IL-18Rα and VEGF-R2, decreased expression of cellular retinol-binding protein-1)	Shimizu et al,[31] Reynhaert et al,[34] Schimitt-Graff et al[35]

Irrespective on their mechanism of activation myofibroblasts seem to migrate towards cancer cell colony even at this very early stage of the metastasis process. As reported by Olaso et al,[24,25] this initiates the development of a stromal support among cancer cells of the emerging small avascular foci. There is a correlation between micrometastasis size and intratumoral density of myofibroblasts, suggesting that factors released from avascular micrometastases induce intratumoral recruitment of additional hepatic stellate cell-derived myofibroblasts. Confirmation of the hepatic stellate cell contribution to tumor-associated myofibroblasts comes from their co-expression of glial fibrillary acidic protein as reported by Olaso et al,[25] a genuine marker of hepatic stellate cells, as shown by Niki et al.[40] No LSECs are usually present in metastatic foci at so early stages of the process.

These histological conclusions have been further analyzed through in vitro models of heterotypic cellular interaction between primary cultured sinusoidal cells and highly-metastatic murine and human cancer cell lines. On the one hand, we have recently observed (Vidal-Vanaclocha et al, unpublished data) that conditioned media obtained from various murine (B16 melanoma, 51b colon carcinoma, C26 colon carcinoma, LY2-head&neck squamous carcinoma) and human (HT-29 carcinoma cells, A375 melanoma) cancer cell lines induced proinflammatory cytokine (TNFα, IL-1β, IL-18, IL-6, etc), PGE-2 and soluble ICAM-1 release from primary cultured murine and human LSECs, respectively. In turn, conditioned medium from tumor-activated murine LSECs also significantly increased chemotaxis, proliferation and matrix metalloprotease-2 release from freshly isolated murine hepatic stellate cells. Interestingly, this was induced by VEGF-dependent IL-18 in the case of stellate cell response to conditioned medium from LSECs pre-activated by murine and human melanoma-derived factors; and it was COX2-dependent in the case of hepatic stellate cell response to conditioned medium from LSECs pre-activated by C26 colon carcinoma-derived factors (Vidal-Vanaclocha et al, unpublished data).

Consistent with in situ observations, the conditioned medium from several human and murine tumor cell lines also directly promoted hepatic stellate cell transdifferentiation into myofibroblasts and increased chemotaxis, proliferation and matrix metalloproteinase-2 release from freshly isolated murine and human hepatic stellate cells, as reported by Olaso et al,[24] Faouzi et al,[26] Shimizu et al.[31] Again, the mechanism was partially VEGF-mediated and COX-2-dependent for several melanoma and carcinoma cell lines. Therefore, from an early stage of the hepatic metastasis process cancer cells can induce hepatic stellate cell transdifferentiation into myofibroblasts directly and/or indirectly via sinusoidal endothelial cell-derived factors of hepatic stellate cells (Fig. 2A).

These redundant mechanisms of hepatic stellate cell stimulation may account for its high ratio per cancer cell during early stages of metastasis. However, histological observation of a large number of avascular micrometastases from different experimental cancer types has shown that occurrence of myofibroblasts at metastatic implantation sites is heterogeneous, suggesting that some cancer cells are unable to induce hepatic stellate cell transdifferentiation into myofibroblasts. Cancer cell proliferation in the absence of a myofibroblastic stroma is still possible, although the resulting metastatic tissue develops necrotic areas very soon. Thus, it is tempting to speculate that myofibroblast recruitment is a prerequisite for the successful development of metastasis in the liver. This would be in agreement with the concept on the metastasis unefficiency of cancer cells at different stages of the metastasis process, postulated by Weiss.[41] Thus, only some specific cancer cell subpopulation would be endowed with the ability induce hepatic stellate cell transdifferentiation. Alternatively (or additionally), they may also release proinflammatory factors for LSECs that would in turn induce stellate cell transdifferentiation. At present, none of these phenotypic properties have been reported among the prometastatic features of cancer cells. Neither exists information on possible biomarker that might serve to identify such cancer cell properties at primary tumor sites. Finally, few studies have yet been done to specifically inhibit hepatic metastasis by targeting tumor-induced hepatic stellate cell transdifferentiation at this early stage of the process (see below).

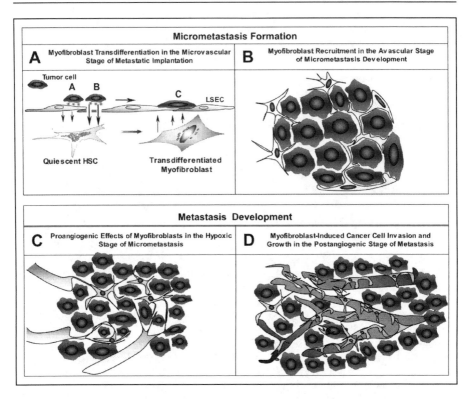

Figure 2. Hepatic stellate cell activation and effects along the hepatic metastasis process. The hepatic colonization process of circulating cancer cells has been schematically represented in 4 consecutive stages with respect to the activation, intratumoral recruitment and tumorigenic actions of perisinusoidal hepatic stellate cells: A) Myofibroblast transdifferentiation in the microvascular stage of metastatic cell implantation; B) Myofibroblast recruitment in the avascular stage of micrometastatic development; C) Proangiogenic effects of myofibroblasts in the prevascularized stage of hypoxic micrometastasis; and D) Myofibroblast-induced cancer cell invasion and growth in the postangiogenic stage of clinically-revelant metastasis.

Hypoxia Induces Proangiogenic Activation of Hepatic Stellate Cell-Derived Myofibroblasts in Avascular Micrometastases: Implications on Intratumoral Endothelial Cell Recruitment and Survival

As above reported, infiltration of tumor-activated myofibroblasts precedes LSEC recruitment into avascular micrometastases. Thereafter, myofibroblasts and endothelial cells co-localized within neoangiogenic capillaries and their numerical densities correlated in developing metastases. Because hypoxic tissue has been identified as a potential source of angiogenic factors within the tumor, we also analyzed the effect of hypoxia on hepatic stellate cell production of LSEC-stimulating factors. As confirmed by pimonidazole staining, hypoxia occurred in hepatic metastases of greater than 300 μm in diameter. However, onset of hepatic stellate cell recruitment occurred in normoxic avascular micrometastases, whereas new intratumoral capillaries are constituted once micrometastases become hypoxic. In vitro, Olaso et al[25] have reported that hypoxia contributes to hepatic stellate cell production of VEGF, which in turn increases endothelial cell migration, reduction of apoptosis, and proliferation.

Using an experimental model of liver cirrhosis, Corpechot et al[42] also showed VEGF production by hypoxic hepatic stellate cell. Thus, their recruitment under normoxic conditions, followed by tumor growth-associated hypoxia, may constitute two synergistic stimuli towards intratumoral migration and survival of LSECs to create a pro-angiogenic microenvironment during the transition of liver metastasis from an avascular to a vascular stage (Fig. 2B,C).

Structural Relationships between Myofibroblastic and Neo-Angiogenic Patterns of Developing Hepatic Metastasis

Consistent with hepatic implantation sites of cancer cells, onset of metastatic growth occurs either from sinusoidal areas or from portal tracts. As reported by Solaun et al[43] and by Olaso et al,[25] two predominant stromal patterns are evident according to expression of α-smooth muscle actin by myofibroblast-like cells: sinusoid-associated metastases, which contain infiltrating, but not encapsulating, reticularly-arranged myofibroblasts; and portal tract-associated metastases, which are incompletely encapsulated, but not infiltrated, by fibrous tract-arranged myofibroblasts.

Based on reticulin staining, the liver architecture is preserved in myofibroblast-infiltrated metastases because invasive cancer cells co-opt the supportive fibrillar network of sinusoids, and, thus, the limit between tumor and normal tissue is ill-defined. In contrast, the reticulin network is not conserved within myofibroblast-encapsulated metastases, because the enlarging mass of cancer cells compresses surrounding parenchyma and generates the formation of tumor lobules delineated by desmoplastic stroma.

These two hepatic micrometastasis variants also are recognized through their microvascular connection and angiogenic pattern, as revealed by CD31/CD34 immunohistochemistry. Micrometastases surrounded by sinusoids contain an internal network of CD31/CD34-expressing sinusoidal-like capillaries associated with myofibroblasts. In contrast, micrometastases located in the vicinity of portal tracts are surrounded by a discontinuous myofibroblast layer, which penetrated the metastatic tissue together with recruited endothelial cells to form long CD31/CD34-expressing angiogenic tracts. Therefore, endothelial cells from sinusoidal vessels might contribute to vascularization of sinusoidal-type metastases. This was not the case for portal-type metastases, the endothelial recruitment of which may derive from portal tract vessels (Fig. 3).

The precise mechanisms underlying different stromal arrangements for sinusoidal and portal-type metastases remain unknown; one possibility is that hepatic stellate cell represent the main source of myofibroblasts for sinusoidal-type metastases, whereas portal tract fibroblasts constitute the stromal support of portal-type metastases. This possibility is supported by the predominant expression of glial fibrillary acidic protein by myofibroblasts located in sinusoidal-type metastases. However, it is also possible that the myofibroblast-inducing activities exerted by cancer cells is dependent on the site of cancer cell implantation or even on its specific capability to activate hepatic stellate cell transdifferentiation into myofibroblasts. LSEC migration only occurred towards avascular micrometastases containing a high density of activated hepatic stellate cell and not towards metastases not containing hepatic stellate cells. Both activated hepatic stellate cell and LSECs colocalized, and their densities consistently correlated with the development of well-vascularized metastases. Conversely, neither cancer cell growth nor LSEC recruitment occurred in hepatic stellate cell-free micrometastases. Interestingly, LSEC recruitment also follows the penetration of hepatic stellate cells into hepatocyte clusters of regenerating liver, as reported by Martinez-Hernandez and Amenta.[44] This physiologic mechanism of tissue reconstitution may also account for the recruitment of LSECs into micrometastases containing activated hepatic stellate cells. Therefore, tumor-activated hepatic stellate cells may promote blood delivery to liver metastasis by triggering the onset of angiogenesis in avascular micrometastases and, then, by supporting their progressive vascularization. Not surprisingly, hepatic stellate cells exhibit pericyte-like functions under physiologic and pathologic conditions as reported by Wake et al[27] and Rockey.[45]

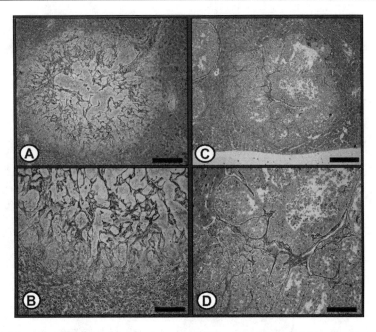

Figure 3. Histochemical study of angiogenic patterns differentiated in sinusoidal-type and portal-type 51b-colon carcinoma hepatic metastases. Reticulin staining was performed according to Gordon-Sweets silver impregnation technique. A) Reticulin histochemistry stained fibrilar matrix network supporting the rich angiogenic microvasculature developed in the sinusoidal-type metastasis. Recruited microvessels form concentric interconnections. Liver architecture is not disturbed, and cancer cells co-opt the supportive fibrilar network of the sinusoids. B) Connections between peritumoral sinusoids and metastatic vessels in the interphase between normal tissue and sinusoidal-type metastasis. C) The reticulin network is not conserved within portal-type metastasis. Desmoplastic stroma surrounds and traverses metastasis, facilitating invasion of vascular-type angiogenic vessels. Necrotic areas frequently develop in this metastasis type. D) Structural connection between metastatic stroma and peritumoral vessels from portal tracts. (bar: 100 μm).

Intrametastatic Myofibroblasts Support Metastasis Development via Paracrine Cancer Cell Invasion and Proliferation-Stimulating Factors

Tumor-associated desmoplastic reaction is a host defense response possibly designed to regulate the developing tumor. However, tumor stroma promotes the growth of the tumor mass as a whole, as above reviewed. It is possible that cancer cells within early lesions undergo additional selection events toward those with an ability to co-opt the new stroma for their own growth advantage. In particular, cancer cells may acquire novel means of heterotypic signaling to myofibroblasts and other stromal cells. Acquisition of such signaling would be expected to provide a growth advantage to the cancer cells and enable them to become invasive. For example, tumor-activated hepatic stellate cell-derived factor(s) promote tumor cell proliferation and migration, as showed by Olaso et al[24] and Shimizu et al.[31] This migration-stimulating activity seems to be partially mediated by platelet-derived growth factor-AB, hepatocyte growth factor and transforming growth factor-β released by hepatic stellate cells. This is cell type-specific, since autologous rat skin fibroblasts did not alter the migration rate of B16 melanoma cells. Therefore, together with direct proangiogenic effects, tumor-associated liver myofibroblasts may also support metastasis development via paracrine cancer cell invasion and proliferation-stimulating factors (Fig. 2D).

Targeting Tumor-Associated Myofibroblasts as a Novel Approach to Anti-Tumor Treatment in the Liver

In situ and in vitro studies suggest that hepatic stellate cell-derived myofibroblasts constitute a key therapeutic target during tumor development. However, their multiple roles along the hepatic metastasis process reveal several therapeutic targeting possibilities against the development and effects of these cells in the cancer microenvironment: First, therapies designed to prevent the myofibroblast transdifferentiation process of tumor-activated hepatic stellate cells; second, therapies to inhibit intratumoral recruitment (chemotactic migration) and multiplication (proliferation) of hepatic stellate cell-derived myofibroblasts; third, therapies to inhibit proangiogenic effects of tumor-associated myofibroblasts on endothelial cells; and fourth, therapies to inhibit release and/or effects of myofibroblast-derived paracrine factors stimulating cancer cell invasion and proliferation.

On this basis, we[46] recently proposed the use of tumor-activated hepatic stellate cells as target for the preclinical testing of anti-angiogenic drugs against hepatic tumor development. In particular, we proposed a two-step study protocol for the screening of anti-angiogenic drugs: (1) In situ analyses on hepatic stellate cell recruitment and co-localization with neo-angiogenic tracts (perivascular coverage) along liver tumor development. (2) In vitro analyses on pro-angiogenic effects of tumor-activated hepatic stellate cells and functional assessment of their effects on primary cultured LSECs. Additional bioassays are currently in development to modelate other functional interactions between tumor-associated hepatic myofibroblasts and cancer cells.

Based on the increased awareness of the important role played by hepatic tumor-associated myofibroblasts in the metastasis process, we and other groups have studied the localization and phenotypic status of hepatic myofibroblasts in the hepatic tumor tissue of animals receiving anti-neoplastic agents. Results demonstrate a statistically significant decrease in the number of tumor-associated hepatic myofibroblasts in treated animals as compared to controls, suggesting that some drugs may exert their anti-tumoral effects in part through inhibitory actions on tumor-activated hepatic stellate cells. For example, a decreased pericytic coverage of angiogenic vessels from colon cancer metastases to liver has been correlated with anti-tumoral effects of the anti-angiogenic tyrosine-kinase inhibitor Su6668, as reported by Shaheen et al.[47] Integrin αV/β3 and αV/β5 inhibitor S247 also diminished intratumoral angiogenesis with a reduction in pericyte coverage as reported by Reinmuth et al.[48] TNP470—a very popular semisynthetic analoge of fumagillin with demonstrated ability to inhibit in vitro HSC transdifferentiation, as reported by Wang et al[49]—significantly decreased liver tumor angiogenesis, as reported by Kinoshita et al[50] and intrametastatic recruitment of hepatic myofibroblasts in vivo (Vidal-Vanaclocha et al, unpublished data). Endostatin—a potent antiangiogenic compound that targets sinusoidal-type liver metastasis as reported by Solaun et al[43]—inhibits tumor-induced hepatic stellate cell migration in vitro and intrametastatic recruitment of hepatic myofibroblasts in vivo (Vidal-Vanaclocha et al, unpublished data). Resveratrol—a natural antioxidant product from grapes, with cyclooxygenase (COX) inhibitory properties and capable of in vitro deactivating human hepatic stellate cells, as reported by Godichaud et al[51]— inhibits prometastatic and proangiogenic activities by tumor-associated myofibroblasts, both in vitro and in vivo (Vidal-Vanaclocha et al, unpublished data). More interestingly, COX-2 inhibitors Celecoxib and Rofecoxib had same effects as Resveratrol, as reported by Wei et al[52] and by Renwick et al.[53] Sodium salicylate—a hypoxia inducible factor-1 α inhibitor that reduces experimental liver tumors as reported by Yang et al[54]—reduced VEGF expression and intratumoral recruitment of hepatic myofibroblasts in vivo. Adaphostin—a tyrosine kinase inhibitor with proapoptotic effects on cancer cells—inhibits VEGF production from tumor-activated hepatic stellate cells in vitro (Vidal-Vanaclocha et al, unpublished data). Another proapoptotic agent— heat shok protein-90 (HSP-90) inhibitor 17-DMA geldanamycin— also significantly decreased intrametastatic recruitment of hepatic myofibroblasts in vivo (Vidal-Vanaclocha et al, unpublished data). Finally, recombinant human IL-18 binding protein inhibits migration of

Table 2. *Anti-tumor therapeutic agents with potential targeting effects on tumor-activated hepatic myofibroblasts during experimental tumor development in the liver*

Compound	Action	Pharmaceutical Mechanism	In Vitro Effect	In Vivo Effect on HSCs on Hepatic Myofibroblasts	References Tumor-Associated
Su6668 (TSU-68)	Inhibitor of receptor tyrosine kinases	Anti-angiogenic agent and tumor growth inhibitor	Not tested	Inhibits pericyte coverage in intratumoral vessels	Shaheen et al[47]
S247	Peptidomimetics of the ligand sequence arg-gly-asp (RGD)	Integrin $\alpha V\beta 3$ and $\alpha V\beta 5$ antagonist	Not tested	Inhibits pericyte coverage in intratumoral vessels	Reinmuth et al[48]
TNP470	Analogue of Fumagillin derivative	Anti-angiogenic agent and tumor growth inhibitor	Inhibits proliferation and activation	Inhibits intratumoral recruitment	Wang et al,[49] Kinoshita et al,[50] Vidal-Vanaclocha et al (unpublished data)
Endostatin	Unknown	Anti-angiogenic agent and tumor growth inhibitor	Inhibits VEGF-induced migration	Inhibits intratumoral recruitment	Solaun et al[43]
Resveratrol	Natural antioxidant and COX inhibitor	Anti-inflammatory and pro-apoptotic agent	Inhibits tumor-induced proliferation and migration	Inhibits intratumoral recruitment	Vidal-Vanaclocha et al (unpublished data), Godichaud[51]
Celecoxib and Rofecoxib	COX-2 inhibitors	Anti-inflammatory and anti-angiogenic agent	Inhibits tumor-induced proliferation and migration	Inhibits intratumoral recruitment	Vidal-Vanaclocha et al (unpublished data), Wei et al,[52] Fenwick et al[53]
Sodium salicilate	HIF-1α transcription blocker	Anti-inflammatory agent	Not tested	Inhibits intratumoral recruitment	Yang ZF et al[54]
Adaphostin (NSC-680410)	Tyrosine kinase inhibitor	Pro-apoptotic agent	Inhibits VEGF secretion from tumor-activated HSCs	Not tested	Vidal-Vanaclocha et al (unpublished data)
17DMA-Geldanamycin (NSC-707545)	Heat shock protein 90 inhibitor	Pro-apoptotic agent	Not tested	Inhibits intratumoral recruitment	Vidal-Vanaclocha et al (unpublished data)
IL-18 binding protein	IL-18 inhibitor	Anti-inflammatory agent	Inhibits tumor-induced migration	Inhibits intratumoral recruitment	Vidal-Vanaclocha et al (unpublished data)

Note: HSC: Hepatic Stellate Cell

tumor-activated human hepatic stellate cells in vitro and intrametastatic recruitment of hepatic myofibroblasts in vivo (Vidal-Vanaclocha et al, unpublished data). Results from these experimental therapeutic assays have been summarized in Table 2.

The idea of targeting the heterotypic interactions that stimulate cancer cell growth with new types of anti-cancer therapies offers some distinct advantages. Cancer-stromal cell interactions may occur in a similar fashion in a wide range of cancer cell types. Targeting the stromal cells rather than the cancer cells themselves may offer more effective responses since the stromal cells appear to have a normal, relatively stable genetic constitution in contrast to diverse and unstable genomes of cancer cells. However, therapeutic targeting of the stromal cells or certain heterotypic interactions may serve just to arrest cancer development and not be effective at causing cancer cell death. Therefore, targeting stromal cells or heterotypic interactions would have to be combined with other therapies that can be cytotoxic to the cancer cells. Still, before such therapies can be envisaged, a better understanding of the precise role of these interactions in hepatic metastasis development is needed. We can anticipate that interest in this area will continue to grow in the coming years.

Conclusions

Hepatic stellate cells transdifferentiate into activated myofibroblasts in the context of several pathophysiologic conditions of human liver. A significant amount of myofibroblasts are also induced and recruited by cancer microenvironment during primary and secondary (metastatic) cancer development in the liver. In experimental models, the main source of tumor-associated myofibroblasts is the hepatic stellate cell, which transdifferentiates into myofibroblasts in response to paracrine factors released by both cancer cells and cancer-activated LSECs and possibly Kupffer cells. Myofibroblasts are already present in the avascular growth stage of developing hepatic metastasis prior to angiogenic endothelial cell recruitment. Further transdifferentiation of myofibroblasts and their intratumor recruitment are supported by intratumor hypoxia. In turn, myofibroblasts may function as supporting stroma for tumor neoangiogenesis. In replacement-type cancer growth, the reticular arrangement of tumor-associated myofibroblasts correlates with a sinusoidal-type angiogenic pattern in metastases. In pushing-type cancer growth, fibrous tract–forming myofibroblasts are located in the peritumoral areas and support a portal-type angiogenic pattern. Finally, tumor-associated myofibroblasts support cancer development via paracrine tumor cell invasion and proliferation-stimulating factors. Therefore, the capability of cancer cells to activate and collaborate with hepatic stellate cells and derived myofibroblasts along tumor progression may represent an unexplored phenotypic property of those cancer cells able to implantate and growth in the liver. On the other hand, several antitumor and anti-angiogenic agents decrease number of tumor-associated hepatic myofibroblasts, suggesting that their anti-tumor effects may occur in part through inhibitory actions on recruitment and tumorigenic effects of these cells. Thus, future studies should be done for better understanding activation, regulation and specific roles of hepatic stellate cell-derived myofibroblasts in the tumor microenvironment. This information may provide new tools for predicting cancer recurrence in liver and may be the basis for target-oriented therapeutic innovation.

References

1. Cheng JD, Weiner LM. Tumors and their microenvironments: Tilling the soil. Clinical Cancer Research 2003; 9:1590-1595.
2. Picard O, Rolland Y, Poupon F. Fibroblast-dependent tumorigenicity of cells in nude mice: Implications for implantation of metastasis. Cancer Res 1986; 46:3290-94.
3. Chung L. Fibroblasts are critical determinants in prostatic cancer growth and dissemination. Cancer Metast Rev 1991; 10:263-274.
4. Grégoire M, Lieubeau B. The role of fibroblasts in tumor behaviour. Cancer Metast Rev 1995; 14:339-350.
5. Elenbaas B, Weinberg RA. Heterotypic Signaling between Epithelial Tumor Cells and Fibroblasts in Carcinoma Formation. Exp Cell Res 2001; 264:169–184.

6. Schürch W, Seemayer Lagacé R et al. Stromal myofibroblasts in primary invasive and metastatic carcinomas. Virchow Arch (Pathol Anat) 1981; 391:125-139.

7. Cornil I, Theodorescu D, Man S et al. Fibroblast cell interactions with human melanoma cells affect tumor cell growth as a function of tumor progression. Proc Natl Acad Sci USA 1991; 88:6028-6032.

8. The International Cancer Microenvironment Society web site (http://cancermicroenvironment.tau.ac.il)

9. Musso O, Theret N Heljasvaara R et al. In situ detection of matrix metalloproteinase-2 (MMP2) and the metalloproteinase inhibitor TIMP2 transcripts in human primary hepatocellular carcinoma and in liver metastasis. J Hepatol 1997; 26:593-605.

10. Theret N, Musso O, Turlin B et al. Increased extracellular matrix remodeling is associated with tumor progresssion in human hepatocellular carcinoma. Hepatology 2001; 34:82-86.

11. Silzle T, Randolph GJ, Kreutz M, et al. The fibroblast: sentinel cell and local immune modulator in tumor tissue. Int J Cancer 2004; 108:173-80.

12. Nakagawa H, Liyanarachchi S, Davuluri RV, et al. Role of cancer-associated stromal fibroblasts in metastatic colon cancer to the liver and their expression profiles. Oncogene 2004; 23:7366-77.

13. Paget S. The distribution of secondary growths in cancer of the breast. Lancet 1889; 1:571-573.

14. Fidler IJ. The pathogenesis of cancer metastasis: the 'seed and soil' hypothesis revisited. Nat Rev Cancer. 2003; 3:453-458.

15. Vidal-Vanaclocha F, Alonso-Varona A, Ayala R et al. Coincident implantation, growth and inter-action sites within the liver of cancer and reactive hematopoietic cells. Int J Cancer 1990; 46:267-271.

16. Jessup J, Battle P, Waller H et al. Reactive nitrogen and oxygen radicals formed during hepatic ischemia-reperfusion kill weakly metastatic colorectal cancer cells. Cancer Res 1999; 59:1825-1829.

17. Weiss L. Biomechanical interactions of cancer cells with the microvasculature during hematog-enous metastasis. Cancer Met Rev 1992; 11:227-235.

18. Vidal-Vanaclocha F, Fantuzzi G, Mendoza L et al. IL-18 regulates IL-1beta–dependent hepatic melanoma metastasis via vascular cell adhesion molecule-1. Proc Natl Acad Sci USA 2000; 97:734-739.

19. Wang H, McIntosh A, Hasinoff B et al. B16 melanoma cell arrest in the mouse liver induces nitric oxide release and sinusoidal cytotoxicity: a natural hepatic defense against metastasis. Cancer Res 2000; 60:5862-5869.

20. Vekemans K, Timmers M, Vermijlen D et al. CC531 colon carcinoma cells induce apoptosis in rat hepatic endothelial cells by the Fas/FasL-mediated pathway. Liver Int 2003; 4:283-93.

21. Luo D, Vermijlen D, Kuppen PJ et al. MHC class I expression protects rat colon carcinoma cells from hepatic natural killer cell-mediated apoptosis and cytolysis, by blocking the perforin/granzyme pathway. Comp Hepatol. 2002; 1:2.

22. Jessup JM, Laguinge L, Lin S et al. Carcinoembryonic antigen induction of IL-10 and IL-6 inhib-its hepatic ischemic/reperfusion injury to colorectal carcinoma cells. Int J Cancer 2004; 111:332-337.

23. Cho D, Song H, Kim Y et al. Endogenous Interleukin 18 modulates immune escape of murine melanoma cells by regulating the expression of Fas ligand and reactive oxygen intermediates. Can-cer Res 2000; 60:2703-2709.

24. Olaso E, Santisteban A, Bidaurrazaga J et al, Tumor-dependent activation of rodent hepatic stellate cells during experimental melanoma metastasis. Hepatology 1997; 26:634-42.

25. Olaso E, Salado C, Egilegor E et al. Proangiogenic role of tumor-activated hepatic stellate cells in experimental melanoma metastasis. Hepatology 2003; 37:674-685.

26. Faouzi S, Lepreux S, Bedin C et al. Activation of cultured rat hepatic stellate cells by tumoral hepatocytes. Lab Invest 1999; 79:485-493.

27. Wake K. Cell-cell organization and functions of 'sinusoids' in liver microcirculation system. J Elec-tron Microsc 1999; 48:89-98.

28. Olaso E, Friedman SL. Molecular Regulation of Hepatic Fibrogenesis. J Hepatol 1998; 29:836-847.

29. Direkze NC, Forbes SJ, Brittan M et al. Multiple organ engraftment by bone-marrow-derived myofibroblasts and fibroblasts in bone-marrow-transplanted mice. Stem cell 2003; 200:429-447.

30. Kalluri R, Neilson G. Epithelial-mesenchymal transition and its implications for fibrosis. J Clin Invest 2003; 112:1776-1784.

31. Shimizu S, Yamada N, Sawada T et al. In vivo and in vitro interactions between human colon carcinoma cells and hepatic stellate cells. Jpn J Cancer Res 2000; 91:1285-1295.

32. Le Pabic H, Bonnier D, Wewer UM et al. ADAM12 in human liver cancers: TGF-beta-regulated expression in stellate cells is associated with matrix remodeling. Hepatology 2003; 37:1056-66.

33. Ooi LP, Crawford DH, Gotley DC et al. Evidence that "myofibroblast-like" cells are the cellular source of capsular collagen in hepatocellular carcinoma. J Hepatol 1997; 26:798-807.

34. Reynaert H, Rombouts K, Vandermonde A et al. Expression of somatostatin receptors in normal and cirrhotic human liver and in hepatocellular carcinoma. Gut 2004; 53:1180-9.

35. Schmitt-Graff A, Ertelt V, Allgaier HP et al. Cellular retinol-binding protein-1 in hepatocellular carcinoma correlates with beta-catenin, Ki-67 index, and patient survival. Hepatology 2003; 38:470-80.
36. Thomas P, Hayashi H, Zimmer R et al. Regulation of cytokine production in carcinoembryonic antigen stimulated Kupffer cells by beta-2 adrenergic receptors: implications for hepatic metastasis. Cancer Lett 2004; 209:251-7.
37. Mendoza L, Carrascal T, de Luca M et al. Hydrogen peroxide mediates vascular cell adhesion molecule-1 expression from IL-18-activated hepatic sinusoidal endothelium: Implications for circulating cancer cell arrest in murine liver. Hepatology 2001; 34:298-310.
38. Mendoza L, Valcarcel M, Carrascal T et al. Inhibition of cytokine-induced microvascular arrest of tumor cells by recombinant endostatin prevents experimental hepatic melanoma metastasis. Cancer Res 2004; 64:304-10.
39. Wisse E, De Zanger RB, Charels K et al. The liver sieve: considerations concerning the structure and function of endothelial fenestrae, the sinusoidal wall and the space of Disse. Hepatology 1985; 5:683-92.
40. Niki T, De Bleser PJ, Xu G et al. Comparison of glial fibrillary acidic protein and desmin staining in normal and CCl4-induced fibrotic rat livers. Hepatology 1996; 23:1538-1545.
41. Weiss L. Dynamic aspects of cancer cell populations in metastasis. Am J Pathol 1979; 97:601-8.
42. Corpechot C, Barbu V, Wendun D et al. Hypoxia-induced VEGF and collagen I expression are associated with angiogenesis and fibrogenesis in experimental fibrosis. Heptology 2002; 35:1010-1021.
43. Solaun MS, Mendoza L, De Luca M et al. Endostatin inhibits murine colon carcinoma sinusoidal-type metastases by preferential targeting of hepatic sinusoidal endothelium. Hepatology 2002 35:1104-1116.
44. Martinez-Hernandez A, Amenta PS. The extracellular matrix in hepatic regeneration. FASEB J 1995; 9:1401-1410
45. Rockey DC. Hepatic blood flow regulation by stellate cells in normal and injured liver. Semin Liver Dis 2001; 21:337-349.
46. Olaso E, Vidal-Vanaclocha F. Use of tumor-activated Hepatic Stellate Cell as target for the preclinical testing of anti-angiogenic compounds against hepatic tumor development. In: Buolamwini K, Adjei A, eds. Methods in Molecular Medicine; Novel Anticancer Drug Protocols. Totowa: The Human Press, 2003:79-86.
47. Shaheen RM, Tseng W, Davis DW et al. Tyrosine kinase inhibition of multiple angiogenic growth factors receptors improves survival in mice bearing colon cancer liver metastases by inhibition of endothelial cell survival mechanism. Cancer Res 2001; 61:1464-1468.
48. Reinmuth N, Liu W, Ahmad SA et al. AlphaVbeta3 integrin antagonist S247 decreases colon cancer metastasis and angiogenesis and improves survival in mice. Cancer Res 2003; 63:2079-2087.
49. Wang YQ, Ikeda K, Ikebe T et al. Inhibition of hepatic stellate cell proliferation and activation by the semisynthetic analogue of fumagillin TNP-470 in rats. Hepatology 2000; 32:980-989.
50. Kinoshita S, Hirai R, Yamano T et al. inhibitor TNP-470 can suppress hepatocellular carcinoma growth without retarding liver regeneration after partial hepatectomy. Surg Today 2004; 34:40-46.
51. Godichaud S, Krisa S, Couronne B et al. Deactivation of cultured human liver myofibroblasts by trans-resveratrol, a grapevine-derived polyphenol. Hepatology 2000; 31:922-931.
52. Wei D, Wang L, He Y et al. Celecoxib inhibits VEGF expression in and reduces angiogenesis and metastasis of human pancreatic cancer via suppression of Sp1 transcription factor activity. Cancer Res 2004; 64:2030-2038.
53. Fenwick SW, Toogood GJ, Lodge JP et al. The effect of the selective cyclooxygenase-2 inhibitor rofecoxib on human colorectal cancer liver metastases. Gastroenterology 2003; 125:716-729.
54. Yang ZF, Poon RT, To J et al. The potential role of HIF1-alpha in tumor progression after hypoxia and chemotherapy in hepatocellular carcinoma. Cancer Res 2004; 64:5496-54503.

Matrix Metalloproteinases, Tissue Inhibitors of Metalloproteinase and Matrix Turnover and the Fate of Hepatic Stellate Cells

Aqeel M. Jamil and John P. Iredale*

Abstract

L iver injury is associated with activation of hepatic stellate cells (HSC) to a myofibroblast-like phenotype. In cirrhotic liver injury, activated HSCs are the major source of fibrillar collagens, an excess of which characterise fibrotic matrix. HSCs also have the capacity to remodel this matrix as they express matrix metalloproteinases (MMPs) and their specific inhibitors, the tissue inhibitors of metalloproteinases (TIMPs). Recovery from acute and chronic injury is characterized by apoptosis of the TIMP expressing HSCs thereby relieving the inhibition of matrix degradation. HSC apoptosis is regulated in progressive injury and counterbalances cell proliferation. Apoptosis probably also represents a default pathway for the HSCs resulting from the withdrawal of survival signals after cessation of injury. The survival of activated HSCs in liver injury is dependent on soluble growth factors and cytokines, and on compontents of the fibrotic matrix itself. Additionally, stimulation of death domain receptors expressed on HSCs can precipitate their apoptosis.

Introduction

In many respects, liver fibrosis can be considered to be a model for solid organ wound healing. There is increasing evidence in models derived from other organs and the skin that demonstrate common features in the processes of inflammation, repair and resolution. Specifically the response to tissue damage is associated with activation of myofibroblasts, the secretion of fibrillar collagens to effectively mediate repair and, with withdrawal of the injurious stimulus, resolution. Resolution is characterized by degradation and remodelling of the fibrillar collagens with the restitution of normal architecture.[1,2] In association with this there is reepithelialisation as well as loss of the myofibroblasts through apoptosis.[3] In liver injury, the wound healing myofibroblasts are derived in major part from activated hepatic stellate cells, although there may also be contributions from peri-portal myofibroblasts.[4,5]

With the advent of effective treatments for chronic liver disease, most importantly the development of interferon and other anti-virals for the treatment of chronic viral hepatitis, there is increasing evidence that liver fibrosis is, at least in part, reversible.[6] Models of liver fibrosis in

*Corresponding Author: John P. Iredale—Southampton University School of Medicine, Liver Research Group, IIR, South Block, Southampton General Hospital, Tremona Road, Southampton, SO16 6YD, U.K. Email: jpi@soton.ac.uk

Tissue Repair, Contraction and the Myofibroblast,
edited by Christine Chaponnier, Alexis Desmoulière and Giulio Gabbiani.
©2006 Landes Bioscience and Springer Science+Business Media.

animals have provided key experimental data which identify the events determining resolution.[7] Prominent amongst these is loss of the activated hepatic stellate cells through apoptosis. This has the effect of removing the major source of fibrillar collagen. Increasing evidence indicates that stellate cell apoptosis is determined by an imbalance in the presence of survival factors and pro-apoptotic stimuli. Amongst these, there is evidence for a role for soluble factors providing survival stimuli and critical changes to the matrix providing survival signals.[2,8] In addition, soluble pro-apoptotic factors have been demonstrated to impact on stellate cells which express a variety of receptors for ligands of the TNF receptor super family.[8-10]

A Brief Review of the Role of Activated Stellate Cells/Myofibroblasts in Hepatic Fibrosis

Liver fibrosis can be considered as a paradigm for wound healing elsewhere in the body. In response to injury, virtually regardless of the insult, the hepatic stellate cell, which is normally a noncycling, quiescent vitamin A storing cell lying in the space of Disse, becomes activated to a myofibroblast-like state (the so called 'activated stellate cell').[11-13] When activated these cells express a variety of cytoskeletal markers, including α smooth muscle actin.[10,14] In addition, the cells enter the growth cycle with the result that the number of activated stellate cells present within the space of Disse and ultimately, in more extensive areas of bridging fibrosis, increases. Stellate cell activation is mediated via the impact of soluble factors secreted by facets of the inflammatory response in addition to products released by damaged hepatocytes and critical changes to the sub-cellular matrix. Once activated, the stellate cells exhibit a number of autocrine and paracrine functions, several of which perpetuate the activation state.[15] These include expression of transforming growth factor β-1.[15] Stellate cell expression of type 1 collagen is significantly upregulated whilst concurrently its degradation is inhibited by expression of TIMPs 1 and 2. However stellate cells also express MMPs, including those with collagenase activity demonstrating the latent capability that the liver has for matrix degradation. Therefore changes in phenotype and cell behaviour leading to the laying down of matrix proteins in which a fibrillar matrix critically predominates depend on the balance between these factors.

Previously considered irreversible, there are extensive (albeit anecdotal or numerically small clinical trials) which have documented an improvement in overall liver fibrosis as a result of the effective treatment of underlying liver disorders. These examples include venesection in haemochromatosis and effective immunosuppression in autoimmune chronic active hepatitis.[16] With the advent of effective anti-viral treatments, however, the first evidence based on large scale studies is available and is providing compelling evidence for at least partial reversibility of fibrotic change in successfully treated patients in whom viral eradication occurs.[6,16] It is important to note, however, that evidence for a reversal of cirrhotic change is as yet incomplete. Indeed compelling histological evidence for reversal of cirrhosis has yet to be demonstrated. Moreover, animal models of advanced cirrhosis do not demonstate the complete resolution observed in models of fibrosis.[17,18] In each of these examples, resolution may take months or years but the improvement in overall histology and the net loss of fibrotic tissue must, by definition, be associated with a net loss of activated hepatic stellate cells. An alternative view is that there may be a change in activation status of the hepatic stellate cells. However, in none of these examples is there evidence of increased numbers of quiescent hepatic stellate cells present in the recovered liver. Furthermore, there is good evidence for resolution of injury being associated with the loss of hepatic stellate cells in the acute setting. Following paracetamol (acetaminophen) injury, in areas of necrosis and inflammation, stellate cells become activated and α smooth muscle actin positive. Successful resolution following conservative treatment was associated with a return to normal histological appearance with a loss of these actin expressing cells. Therefore, this study provides direct evidence that resolution of injury is associated with a reduction in the number of α smooth muscle positive myofibroblast cells.[1]

Animal models complement these human models of resolution of liver disease. Exposure of rodents to chronic CCl₄ intoxication results in the development of fibrosis and ultimately cirrhosis over a period of four to twelve weeks. The resulting scars link vascular structures and are populated by large numbers of activated stellate cells which have proliferated in response to the injurious and inflammatory stimuli. Following induction of an advanced fibrosis or early cirrhosis (4 to 8 weeks of CCl₄ intoxication) spontaneous resolution occurs with a restitution of normal liver architecture and loss of the activated stellate cells over a period of 28 to 168 days. This process of resolution is accompanied by a decrease in the hepatic expression of TIMP-1 and an increase in overall hepatic collagenase activity.[7]

A similar change can be demonstrated following the induction of bile duct ligation induced fibrosis if the ligated bile duct is successfully reanastomosed to the jejunum with resulting biliary decompression.[2,19] In each of these models, the loss of α smooth muscle actin positive activated stellate cells is mediated by apoptosis. Thus, resolution is characterised not only by changes in the pattern of matrix degradation but by apoptosis of myofibroblast-like activated hepatic stellate cells. This process serves the dual function of removing the major matrix-producing cell whilst at the same time removing the cells that are expressing TIMPs and thus inhibiting matrix degradation. We have subsequently gone on to demonstrate the beneficial effects of removing activated hepatic stellate cells by inducing apoptosis using the fungal metabolite gliotoxin. When this toxin is given to rats during chronic CCl₄ induced injury, apoptosis of stellate cells rapidly ensues with significant decrease in the overall number of activated hepatic stellate cells and a reduction in the width of fibrotic septa.[2,20] This work has recently been corroborated in several other laboratories.[2,21-23] In addition, in data as yet unpublished, we have gone on to demonstrate that induction of activated stellate cell apoptosis via NGF stimulation (see below) also results in amelioration of the fibrotic response.

The Regulation of Hepatic Stellate Cell Apoptosis

Apoptosis is a major regulatory mechanism active in mammalian tissues which removes unwanted cells when they become too numerous, redundant or potentially damaging, e.g., oncogenesis. There is evidence for a constant background of apoptosis in activated hepatic stellate cells during liver injury and it is likely that this is a major mechanism regulating overall stellate cell numbers. In common with other cells, stellate cells demonstrate susceptibility to apoptosis induced via two basic intracellular pathways. These include stimulation of specific cell surface receptors which carry a so-called death domain, e.g., Fas (CD-95, AP0-1). Exposure of the cell to the relevant ligand results in activation of the death receptor and activation of the intracellular cascade which may result in apoptosis. Although activation of the caspase cascade leading to the activation of caspase-3 is frequently an effective means of inducing apoptosis, this pathway is, nevertheless susceptible to modification and inhibition.[24] The second major pathway involves the stability and integrity of the mitochondrial membrane. Pro and anti-apoptotic proteins, present in the mitochondrial membrane, (e.g., BCL-2 family members) are imbalanced. When pro-apoptotic proteins predominate and are allowed to homo-dimerise, cytochrome C is released from the mitochondrion which complexes with APAF-1 resulting in activation of specific caspases and apoptotic death of the cells. The balance of mitochondrial membrane proteins is regulated in major part via specific signals received by the cell, including those derived from soluble factors and cell-cell and cell-matrix interactions.[25]

The stellate cell fits very nicely the model proposed by Raff[26,27] in which a cell is imminently prone to undergo apoptosis as the default position but this process is forestalled by the presence of survival signals. These signals may be derived from the cellular environment in the form of soluble growth factors and cytokines, or in the example of hepatic stellate cell from the matrix (see below). During the resolution of injury, apoptosis will ensue when these survival factors fall below a critical level and the balance of pro and anti-apoptotic factors in the cells shift.

Soluble Cytokines and Survival Factors in the Regulation of Stellate Cell Apoptosis

Several cytokines and growth factors released during liver cell injury may impact on stellate cells apoptosis. These include IGF-1 released in an autocrine manner by damaged hepatocytes and by stellate cells.[28,29] IGF-1 is a powerful survival factor for stellate cells and may act in concert with other soluble factors.[30-33] There is published evidence to suggest that tumour necrosis factor (TGF)-β1 may also regulate stellate cell survival, as may TNFα, although the latter may mediate its effect via fas/fas ligand (see below).[34]

The Role of TNF Receptor Super Family Members in Mediating Stellate Cell Apoptosis and Survival

Several members of the TNF receptor super family bear a so-called death domain and are caspase activating, when stimulated with the appropriate ligands. Stellate cells have been described to express fas (CD-95 APO-1) and to respond to fas ligand by undergoing apoptosis.[8,35] Stellate cells have also been reported to express fas ligand itself, a product which is cleaved by metalloproteinases to yield a soluble cell signal. Thus, it is possible that stellate regulate their own survival via this autocrine loop.[9,34,36] We, and others, have also described the expression of a further member of the TNF receptor super family, low affinity nerve growth factor receptor (LANGFR) or p75. LANGFR/p75 is expressed by activated stellate cells which undergo apoptosis in response to nerve growth factor.[8-10,37-41] Subsequent studies using a model of self-limiting fibrotic injury have indicated that during resolution and particularly whilst undergoing proliferation, hepatocytes are a potent source of NGF. Thus we have demonstrated a potential paracrine loop in which the injured liver by regenerating hepatic epithelium affects apoptosis of hepatic stellate cells via NGF.[42]

Matrix Stability and the Role of Tissue Inhibitor of Metalloproteinases in Mediating Stellate Cell Survival

Because of the close correlation between TIMP expression within the liver and survival of hepatic stellate cells, we have directly addressed whether TIMP-1 may act as survival factor for hepatic stellate cells. Previously, evidence from oncological studies indicated that TIMP-1 may mediate survival of specific tumours, although intriguingly these effects appeared independent of MMP inhibitory activity.[43] In tissue culture models of activated hepatic stellate cells, TIMP-1 acts as a potent pro survival factor for stellate cells.[44] By using TIMP-1 bearing a critical mutation (resulting in a failure to bind and then inhibit MMPs) we have demonstrated that MMP inhibitory activity is necessary for the anti-apoptotic effect of TIMP-1 for hepatic stellate cells.[45] This work has highlighted a potential link between MMP activity and stellate cell apoptosis. Specifically it poses the question, what are the critical MMPs and what are the MMP substrates that these MMPs are acting upon that impact on activated stellate cell survival?

We have gone on to examine two potential substrates which may act as survival factors for stellate cells. N-cadherin has been shown to mediate cell-cell contacts in fibroblasts and promote survival in 3T3 cells.[46] In addition, MMP inhibitors have been shown to upregulate N-cadherin function and promote survival in fibroblasts. In unpublished studies we have demonstrated that activated hepatic stellate cells express N-cadherin and antibody-mediated blockade of N-cadherin promotes stellate cell apoptosis. Analysis of N-cadherin structure by Western blotting, during stellate cell apoptosis, indicates that N-cadherin cleavage is an early apoptotic event. Moreover, immunostaining demonstrates loss of intact N-cadherin from the cell surface early in stellate cell apoptosis. Addition of wild-type active TIMP-1 results in a reduction of stellate cell apoptosis, both in response to a stress such as serum deprivation and in response to a specific apoptogen such as gliotoxin or cyclohexamide. This is accompanied by a decrease in the cleavage of N-cadherin detected by Western analysis. In contrast, parallel experiments undertaken with the mutated TIMP-1 demonstrate persistent cleavage and enhanced level of apoptosis. We have identified that a synthetic MMP-2 inhibitor will also prevent a cleavage of N-cadherin suggesting that this enzyme, which is released and activated by apoptotic hepatic stellate cells may be responsible for cleavage of N-cadherin. Further experiments in which recombinant MMP-2 have been added to stellate cells demonstrate both cleavage of N-cadherin and enhanced apoptosis.[43,44,47]

A further substrate, likely to impact on stellate cell apoptosis is collagen-1. There is an extensive literature suggesting that matrix may dramatically regulate stellate cell function.[21-23,48] Indeed we and others have recently demonstrated that stellate cells can be reverted to quiescence by culture on a model basement membrane like matrix Engelbreth-Holm-Swarm (EHS). In contrast stellate cells plated on to type 1 collagen maintain α smooth muscle actin positivity and demonstrate a propensity to apoptosis on withdrawal of growth factors. Intriguingly, expression of the anti-apoptotic protein, Bcl-2 appears throughout regulated by sub-cellular matrix. Indeed, plating hepatic stellate cells onto EHS matrix results in enhanced expression of Bcl-2 in contrast to cells plated on to Type 1 collagen.[49-51] Because of the tight correlation between Type 1 collagen degradation and stellate cell apoptosis in vivo, we have studied the role of collagen-1 as a survival factor using a transgenic in vivo model. The rr collagen mouse contains a targeted mutation of the collagen-1 gene which results in the secretion of a gene product resistant to cleavage mediated by MMP-2, MMP-8 or MMP-13. Using these mice in comparison to wild type counterparts, we have induced reversible fibrosis by giving 8 weeks of CCl_4. Animals were then allowed to recover for periods of up to 28 days. Wild type animals demonstrated a significant resolution of the fibrotic change in association with which stellate cell apoptosis resulted in clearance of the α smooth muscle positive activated hepatic stellate cells. In contrast, there was persistence of fibrotic change in the mutant animals, indicating a failure of degradation of the collagenase resistant collagen-1, in association with which activated stellate cells persisted within the liver and there was a failure of hepatocellular regeneration.[51]

This study provides very cogent evidence that collagen-1 may be a specific survival factor for stellate cells. We have gone on to analyze the integrin which bind to stellate cells and have demonstrated that wild type collagen-1 promotes the activated phenotype and proliferation of stellate cells via interaction with β-1 integrins and $\alpha_V\beta_3$. However, as noted above, this phenotype is associated with a propensity to apoptosis and blockade, particularly of $\alpha_V\beta_3$ is associated, not only with a decrease in proliferation, but a significant increase in apoptosis.[51,52]

Taken together, this provides cogent evidence for a link between the persistence of collagen-1 rich matrix and stellate cell survival. In contrast, with effective degradation of collagen-1, stellate cells are rendered susceptible to apoptotic stimuli or undergo a critical change in the balance of survival and pro-apoptotic factors resulting in their entering the apoptotic pathway.

These latter observations have led us to postulate that critical changes, involving modification of the matrix, may, therefore, result in persistent fibrosis and potentially persistence of hepatic stellate cells, even after withdrawal of injurious stimuli. Advanced cirrhosis is characterised by the presence of fibrotic bands linking vascular structures and regenerative nodules of hepatic parenchyma. Whilst there is good evidence for a reduction in the overall level of fibrosis following successful anti-viral eradication, there is, as yet, no absolute evidence that cirrhosis will undergo complete resolution. Moreover, studies of human liver disease suggest that in the presence of a micronodular cirrhosis, withdrawal of the injurious stimulus may result in degradation of the least mature matrix and a retreat of the pathological pattern to an attenuated macronodular cirrhosis. To complement these observations and to determine whether matrix modifications impact on reversibility, we have recently developed a further novel of advanced micronodular cirrhosis.[17,18,53,54] In rats, given CCl_4 for 12 weeks, a micronodular cirrhosis results. Although there is a significant improvement over a year of spontaneous recovery, significant fibrosis persists in the form of an attenuated macronodular cirrhosis. Analysis of the persistent fibrosis suggests that these are areas of the most mature fibrosis and are characterised by the presence of elastin and tissue transglutaminase mediated crosslinks. Stellate cells persist in these fibrotic bands although, after protracted recovery, the numbers of α smooth muscle actin positive stellate cells within the bands are limited and the persistent cells express predominantly Glial Fibrillary Acidic Protein and desmin alone.[18]

Summary

A wealth of evidence indicates the activated hepatic stellate cell is central to the pathogenesis of liver fibrosis, being both the major source of matrix, particularly fibrillar collagens, and the major source of the metalloproteinase inhibitors which prevent degradation of that matrix. Other myofibroblast lineages likely to contribute to the hepatic scar and may be derived from stem cells and other endogenous liver myofibroblasts. Current evidence suggests that functionally this population contribute in a broadly similar way to the hepatic scar. Spontaneous recovery from liver fibrosis and the associated resolution of fibrotic change is accompanied by apoptosis of hepatic stellate cells. Our current model shows that activated hepatic stellate cells are prone to apoptosis but that this is forestalled by the presence of survival factors. During recovery, the balance of survival factors versus pro-apoptotic stimuli shifts with the result that the cells initially become prone to apoptosis and ultimately enter the apoptotic pathway as a result of loss of survival factors. There is evidence that these survival factors may be derived from inflammatory cells and injured hepatocytes in addition to critical changes in the matrix and the stabilisation of cell-cell receptors such as N-cadherin. With resolution, there is evidence for expression of pro-apoptotic stimuli including Nerve Growth Factor and with withdrawal of stimuli from injured cells and the inflammatory infiltrate, stellate cell apoptosis ensues. Critical changes in the matrix resulting in the loss of fibrillar collagen, particularly collagen-1, may also then result in stellate cell apoptosis. Stabilisation of the matrix via crosslinking may limit the propensity for histological resolution and functional restitution.

By understanding the processes that regulate hepatic stellate cell apoptosis, we will define the attributes of an effective anti-fibrotic agent and inform the development of future anti-fibrotic strategies.

References

1. Mathews J et al. Non paranchymal cell responses in paracetamol induced liver injury. J Hepatol 1994; 20:537-537.
2. Issa R, Williams E, Trim N et al. Apoptosis of hepatic stellate cells: Involvement in resolution of biliary fibrosis and regulation by soluble growth factors. Gut 2001; 48:548-557.
3. Hautekeete ML, Geerts A. Virchows Arch 1997; 430(3):195-207.
4. Knittel T, Kobold D, Saile B et al. Rat liver myofibroblasts and hepatic stellate cells: Different cell populations of the fibroblast lineage with fibrogenic potential. Gastroenterology 1999; 117:1205-1221.
5. Knittel T, Kobold D, Piscaglia F et al. Localization of liver myofibroblasts and hepatic stellate cells in normal and diseased rat livers: Distinct roles of (myo-)fibroblast subpopulations in hepatic tissue repair. Histochem Cell Biol 1999; 112:387-401.
6. Lau DT, Kleiner DE, Park Y et al. Resolution of chronic delta hepatitis after 12 years of interferon alfa therapy. Gastroenterology 1999; 117:1229-1233.
7. Iredale JP, Benyon RC, Pickering J et al. Mechanisms of spontaneous resolution of rat liver fibrosis. Hepatic stellate cell apoptosis and reduced hepatic expression of metalloproteinase inhibitors. J Clin Invest 1998; 102:538-549.
8. Krammer PH. CD95's deadly mission in the immune system. Nature 2000; 407:789-795.
9. Saile B, Knittel T, Matthes N et al. CD95/CD95L-mediated apoptosis of the hepatic stellate cell. A mechanism terminating uncontrolled hepatic stellate cell proliferation during hepatic tissue repair. Am J Pathol 1997; 151:1265-1272.
10. Vaux DL, Strasser A. The molecular biology of apoptosis. Proc Natl Acad Sci USA 1996; 93:2239-2244.
11. Friedman SL. Seminars in medicine of the Beth Israel Hospital, Boston. The cellular basis of hepatic fibrosis. Mechanisms and treatment strategies. N Engl J Med 1993; 328:1828-1835.

12. Alcolado R, Arthur MJ, Iredale JP. Pathogenesis of liver fibrosis. Clin Sci (Lond) 1997; 92:103-112.
13. Friedman SL, Roll FJ, Boyles J et al. Hepatic lipocytes: The principal collagen-producing cells of normal rat liver. Proc Natl Acad Sci USA 1985; 82:8681-8685.
14. Tanaka Y, Nouchi T, Yamane M et al. Phenotypic modulation in lipocytes in experimental liver fibrosis. J Pathol 1991; 164:273-278.
15. Scharf JG, Schmidt-Sandte W, Pahernik SA et al. Characterization of the insulin-like growth factor axis in a human hepatoma cell line (PLC). Carcinogenesis 1998; 19:2121-2128.
16. Powell LW, Kerr JF. Reversal of "cirrhosis" in idiopathic haemochromatosis following long-term intensive venesection therapy. Australas Ann Med 1970; 19:54-57.
17. Desmet VJ, Roskams T. Cirrhosis reversal: A duel between dogma and myth. J Hepatol 2004; 40:860-867.
18. Issa R, Zhou X, Constandinou CM et al. Spontaneous recovery from micronodular cirrhosis: Evidence for incomplete resolution associated with matrix cross-linking. Gastroenterology 2004; 126:1795-1808.
19. Hammel P, Couvelard A, O'Toole D et al. Regression of liver fibrosis after biliary drainage in patients with chronic pancreatitis and stenosis of the common bile duct. N Engl J Med 2001; 344:418-423.
20. Wright MC, Issa R, Smart DE et al. Gliotoxin stimulates the apoptosis of human and rat hepatic stellate cells and enhances the resolution of liver fibrosis in rats. Gastroenterology 2001; 121:685-698.
21. Orr JG, Leel V, Cameron GA et al. Mechanism of action of the antifibrogenic compound gliotoxin in rat liver cells. Hepatology 2004; 40:232-242.
22. Dekel R, Zvibel I, Brill S et al. Gliotoxin ameliorates development of fibrosis and cirrhosis in a thioacetamide rat model. Dig Dis Sci 2003; 48:1642-1647.
23. Kweon YO, Paik YH, Schnabl B et al. Gliotoxin-mediated apoptosis of activated human hepatic stellate cells. J Hepatol 2003; 39:38-46.
24. Hengartner MO. The biochemistry of apoptosis. Nature 2000; 407:770-776.
25. Savill J, Fadok V. Corpse clearance defines the meaning of cell death. Nature 2000; 407:784-788.
26. Raff MC, Barres BA, Burne JF et al. Programmed cell death and the control of cell survival: Lessons from the nervous system. Science 1993; 262:695-700.
27. Raff MC. Social controls on cell survival and cell death. Nature 1992; 356:397-400.
28. Brenzel A, Gressner AM. Characterization of insulin-like growth factor (IGF)-I-receptor binding sites during in vitro transformation of rat hepatic stellate cells to myofibroblasts. Eur J Clin Chem Clin Biochem 1996; 34:401-409.
29. Pinzani M, Abboud HE, Aron DC. Secretion of insulin-like growth factor-I and binding proteins by rat liver fat-storing cells: Regulatory role of platelet-derived growth factor. Endocrinology 1990; 127:2343-2349.
30. Valentinis B, Reiss K, Baserga R. Insulin-like growth factor-I-mediated survival from anoikis: Role of cell aggregation and focal adhesion kinase. J Cell Physiol 1998; 176:648-657.
31. Resnicoff M, Burgaud JL, Rotman HL et al. Correlation between apoptosis, tumorigenesis, and levels of insulin-like growth factor I receptors. Cancer Res 1995; 55:3739-3741.
32. Resnicoff M, Burgaud JL, Rotman HL et al. Correlation between apoptosis, tumorigenesis, and levels of insulin-like growth factor I receptors. Cancer Res 1995; 55:3739-3741.
33. Mooney A, Jobson T, Bacon R et al. Cytokines promote glomerular mesangial cell survival in vitro by stimulus-dependent inhibition of apoptosis. J Immunol 1997; 159:3949-3960.
34. Saile B, Matthes N, Knittel T et al. Transforming growth factor beta and tumor necrosis factor alpha inhibit both apoptosis and proliferation of activated rat hepatic stellate cells. Hepatology 1999; 30:196-202.
35. Tanaka M, Suda T, Haze K et al. Fas ligand in human serum. Nat Med 1996; 2:317-322.
36. Gong W, Pecci A, Roth S et al. Transformation-dependent susceptibility of rat hepatic stellate cells to apoptosis induced by soluble Fas ligand. Hepatology 1998; 28:492-502.
37. Cleveland JL, Ihle JN. Contenders in FasL/TNF death signaling. Cell 1995; 81:479-482.
38. Dechant G, Barde YA. Signalling through the neurotrophin receptor p75NTR. Curr Opin Neurobiol 1997; 7:413-418.
39 Galle PR, Hofmann WJ, Walczak H et al. Involvement of the CD95 (APO-1/Fas) receptor and ligand in liver damage. J Exp Med 1995; 182:1223-1230.
40. Smith CA, Farrah T, Goodwin RG. The TNF receptor superfamily of cellular and viral proteins: Activation, costimulation, and death. Cell 1994; 76:959-962.
41. Trim N, Morgan S, Evans M et al. Hepatic stellate cells express the low affinity nerve growth factor receptor p75 and undergo apoptosis in response to nerve growth factor stimulation. Am J Pathol 2000; 156:1235-1243.

42. Oakley F, Trim N, Constandinou CM et al. Hepatocytes express nerve growth factor during liver injury: Evidence for paracrine regulation of hepatic stellate cell apoptosis. Am J Pathol 2003; 163:1849-1858.
43. Guedez L, Stetler-Stevenson WG, Wolff L et al. In vitro suppression of programmed cell death of B cells by tissue inhibitor of metalloproteinases-1. J Clin Invest 1998; 102:2002-2010.
44. Murphy FR, Issa R, Zhou X et al. Inhibition of apoptosis of activated hepatic stellate cells by tissue inhibitor of metalloproteinase-1 is mediated via effects on matrix metalloproteinase inhibition: Implications for reversibility of liver fibrosis. J Biol Chem 2002; 277:11069-11076.
45. Murphy FR, Issa R, Zhou X et al. Inhibition of apoptosis of activated hepatic stellate cells by tissue inhibitor of metalloproteinase-1 is mediated via effects on matrix metalloproteinase inhibition: Implications for reversibility of liver fibrosis. J Biol Chem 2002; 277:11069-11076.
46. Luegmayr E, Glantschnig H, Varga F et al. The organization of adherens junctions in mouse osteoblast-like cells (MC3T3-E1) and their modulation by triiodothyronine and 1,25-dihydroxyvitamin D3. Histochem Cell Biol 2000; 113:467-478.
47. Murphy F, Waung J, Collins J et al. N-Cadherin cleavage during activated hepatic stellate cell apoptosis is inhibited by tissue inhibitor of metalloproteinase-1. Comp Hepatol 2004; 3(Suppl 1):S8.
48. Trim N et al. Intact collagen-I inhibits hepatic stellate cell (HSC) activation and promotes persistence of activated HSC in vivo. Hepatology 2004; 32:183.
49. Gaca MDA, Kirella K, IJIRBRC. Extracellular matrix regulates hepatic stellate cell phenotype and survival. J Hepatol 2000; 32-83.
50. Han YP, Zhou L, Wang J et al. Essential role of matrix metalloproteinases in interleukin-1-induced myofibroblastic activation of hepatic stellate cell in collagen. J Biol Chem 2004; 279:4820-4828.
51. Issa R, Zhou X, Trim N et al. Mutation in collagen-1 that confers resistance to the action of collagenase results in failure of recovery from CCl4-induced liver fibrosis, persistence of activated hepatic stellate cells, and diminished hepatocyte regeneration. FASEB J 2003; 17:47-49.
52. Zhou X, Murphy FR, Gehdu N et al. Engagement of alphavbeta3 integrin regulates proliferation and apoptosis of hepatic stellate cells. J Biol Chem 2004; 279:23996-24006.
53. Desmet VJ, Roskams T. Reversal of cirrhosis: Evidence-based medicine? Gastroenterology 2003; 125:629-630.
54. Wanless IR, Nakashima E, Sherman M. Regression of human cirrhosis. Morphologic features and the genesis of incomplete septal cirrhosis. Arch Pathol Lab Med 2000; 124:1599-1607.

CHAPTER 11

Innate Immune Regulation of Lung Injury and Repair

Dianhua Jiang, Jennifer Hodge, Jiurong Liang and Paul W. Noble*

Abstract

Mechanisms that regulate host defense to noninfectious tissue injury are poorly understood. Here we summarize our recent work investigating the role of the innate immune response in regulating the inflammatory and fibrotic response to noninfectious lung injury. We have identified key roles for two cell surface receptors in regulating lung inflammation and fibrosis. CD44 has an essential role in resolving inflammation following noninfectious lung injury. A major function of CD44 in vivo is to clear hyaluronan degradation products that are produced following lung injury. Failure ot clear hyaluronan leads to unremitting inflammation. In contrast, the chemokine receptor CXCR3 has an essential role in limiting the extent of fibrosis following lung injury. This protective effect of CXCR3 in limiting tissue fibroproliferation is mediated, in part, by the innate production of interferon-γ following lung injury. These studies reveal a previously unsuspected role for the innate immune response in regulating inflammation and fibrosis.

Host Responses in Lung Injury

Regulation of tissue injury and repair is a carefully orchestrated host response to eradicate the offending agent and restore tissue integrity. Successful repair of tissue injury requires a coordinated host response to limit the extent of structural cell damage. The mechanisms that regulate the host response to tissue injury are incompletely understood. Abnormalities in the host repair response are associated with a variety of chronic disease states that are characterized by excessive deposition of extracellular matrix resulting in organ fibrosis. The innate response to an insult is the component of the host response that is in place prior to the insult and serves as the first line of defense against injury. For example, following exposure to an infectious pathogen, the initial phase of the host response consists of macrophages recognizing the presence of the pathogen and releasing signals that trigger an inflammatory response. This innate response is mediated by Toll-like receptors (TLRs) on macrophages that orchestrate the recognition of the invading pathogen and initiation of the host inflammatory response with the goal of restoration of tissue integrity.[1,2] In concert with initiating the inflammatory response, the host must also generate signals to minimize the extent of structural cell damage. The ultimate outcome of the host depends on the balance between containment of injury, maintenance of structural cell integrity and activating repair mechanisms. While much is known about the innate immune response to infectious agents, the mechanisms that regulate the host response to noninfectious tissue injury

*Corresponding Author: Paul W. Noble—Section of Pulmonary and Critical Care Medicine, Yale University School of Medicine, TAC 441-C, New Haven, Connecticut 06520-8057, U.S.A. Email: paul.noble@yale.edu

Tissue Repair, Contraction and the Myofibroblast,
edited by Christine Chaponnier, Alexis Desmoulière and Giulio Gabbiani.
©2006 Landes Bioscience and Springer Science+Business Media.

are poorly understood. In particular, the relationship between host defense to noninfectious injury and organ fibrosis has not been explored. This review will focus on recent work from our laboratory in regard to understanding the innate immune response to noninfectious lung injury and effects of host defense on lung inflammation and fibrosis.

Hyaluronan and CD44 in Lung Injury and Repair

A hallmark of tissue injury is increased turnover of extracellular matrix (ECM). A variety of lung diseases are associated with abnormal ECM turnover. Asthma, emphysema and pulmonary fibrosis are three chronic lung diseases that have in common an imbalance between the synthesis and degradation of ECM. In chronic lung diseases, the ECM is modified by the inflammatory milieu and degradation products are generated by oxidants and other mechanisms that take on unique properties not attributable to the precursor molecules. Previous work from our laboratory has shown that failure to remove matrix degradation products from the lung following injury results in the host succumbing to unremitting inflammation.[3] We have focused our studies on hyaluronan (HA). HA is an ECM component that undergoes dynamic regulation during tissue injury and inflammation. HA is a nonsulfated glycosaminoglycan composed of repeating polymeric disaccharides *D*-glucuronic acid and *N*-acetyl glucosamine. HA can exist as both a soluble polymer and noncovalently linked to proteoglycan core proteins. In the lung, the main proteoglycan with HA side chains is versican. Under physiologic conditions HA exists as a high molecular weight polymer ($>10^6$ Da) and undergoes dynamic regulation resulting in accumulation of lower molecular weight species following tissue injury.[3,4] HA has essential functions in vivo. In normal development, absence of HA by genetically targeting deletion of hyaluronan synthase 2 results in embryonic lethality due to severe cardiac and vascular abnormalities.[5] HA degradation products generated in vitro induce the expression of a variety of genes including chemokines, cytokines, growth factors, signal transduction molecules, and adhesion molecules in macrophages, eosinophils, dendritic cells, and fibroblasts, suggesting endogenously generated HA fragments may regulate inflammatory processes.[6-10] CD44 is a type 1 transmembrane glycoprotein that is expressed on most cells and is a major cell surface receptor for HA. The interaction of HA and CD44 has been implicated in the regulation of a variety of biological processes including tumor growth and metastasis, wound healing, T cell recruitment to sites of inflammation, macrophage activation, neutrophil migration and endothelial cell activation. However, unlike the targeted deletion of HAS2, mice with a targeted deletion in CD44 were found to develop normally.[11] In order to investigate the role of HA in lung injury and repair, we examined the inflammatory response in CD44-deficient mice.[3] The model system that we and others have utilized to study noninfectious lung injury and repair is to instill the DNA-damaging chemotherapeutic agent bleomycin sulfate into the tracheas of mice. Bleomycin sulfate causes epithelial cell injury, elicits an acute inflammatory response that usually peaks by day 7. Over the ensuing 14 days, the inflammatory response remits and there is a robust fibroproliferative response that results in the significant deposition of extracellular matrix components such as collagen as well as the transient presence of myofibroblasts. Bleomycin-induced lung injury has been a useful model to study noninfectious lung inflammation and repair. In addition, to the deposition of collagen, an important aspect of the injury and repair response is the accumulation and subsequent clearance of HA. HA production rises several fold following bleomycin treatment and peaks between days 7-10. HA is then cleared from the lung interstitium coincident with the accumulation of collagen. We and others, have shown that HA degradation products accumulate in the lung in the size range (10-500 kDa) that can stimulate inflammatory gene expression by macrophages and other inflammatory cells in vitro. In order to examine the role of HA homeostasis in lung injury and repair we examined the inflammatory response to intratracheal bleomycin sulfate in CD44-deficient mice.[3] We made the unexpected observation that CD44-deficiency led to an increased susceptibility to lung injury with a marked increase in mortality following bleomycin treatment (Fig. 1). Examination of the lung tissue at a time point when the wild type mouse lung had resolved the inflammatory

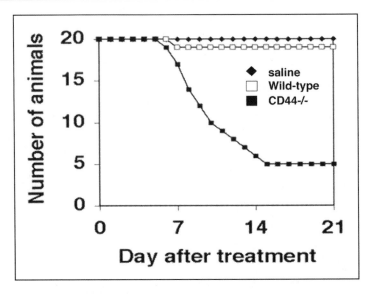

Figure 1. CD44 deficiency leads to increased mortality following bleomycin-induced lung injury. Reproduced with permission from: Teder et al, Science 2002; 296:155-158, ©2002 The American Association for the Advancement of Science.

response, demonstrated an overwhelming accumulation of inflammatory cells in the CD44-deficient mice (Fig. 2). In addition to the inability to resolve the lung inflammation, we also identified a striking abnormality in the deposition of HA. As shown in Figure 3, there is abundant HA accumulation in the injury lungs in the CD44 deficient mice relative to the wild type control mice. CD44-deficient mice were unable to clear HA degradation products from the lung interstitium. The CD44-deficient state was associated with both an inability to resolve lung inflammation and properly remove HA from the lung tissue. Evaluation of the molecular mass of the lung tissue HA revealed major differences between the wild type and CD44-deficient states. CD44-deficiency resulted in the accumulation of both higher and lower molecular mass species than observed in the wild type mice. CD44 thus has a profound role in regulating HA homeostasis in the lung following acute injury. Since CD44 is present on both structural cells (epithelium, fibroblasts, and endothelium) as well and hematopoietic cells we sought to determine which cellular compartment was responsible for HA clearance and resolution of inflammation. We generated chimeric mice by transferring bone marrow from syngeneic wild type mice into sublethally irradiated CD44 deficient mice. We thus reconstituted CD44 positive hematopoietic cells in CD44 deficient mice. When these chimeric mice were then treated with bleomycin we found that the profound increase in mortality was prevented. This reversal in phenotype was accompanied by a restoration of the ability to resolve the inflammatory response and remove HA from the lung interstitium.[3] Collectively, these data suggest that the production of HA following acute tissue injury serves the very important function of initiating the host innate immune response by providing an essential signal to macrophages to produce chemokines that recruit other leukocyte subsets required to debride the tissue injury and begin restoring tissue integrity. In addition, while it appears essential to the host to initiate inflammatory responses when HA degradation products are recognized by macrophages, it is also important that the inflammatory response be successfully resolved. An important component of the resolution of the inflammatory response is the successful removal of HA degradation products. Failure to properly clear these ECM breakdown products results in unremitting inflammation. These data suggest that a previously unrecognized component of the innate immune response is the generation and

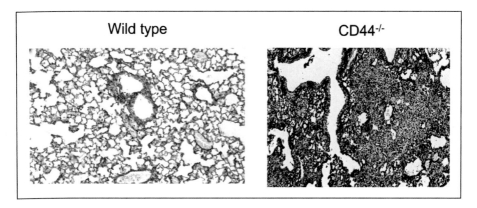

Figure 2. CD44 deficiency leads to unremitting lung inflammation following bleomycin-induced lung injury.

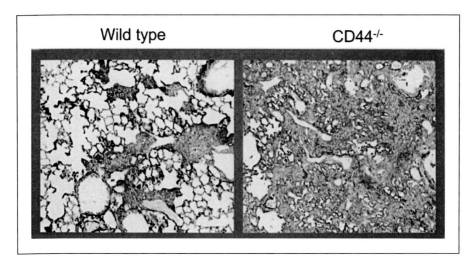

Figure 3. CD44 deficiency leads to impaired clearance of hyaluronan following bleomycin-induced lung injury. Reproduced with permission from: Teder et al, Science 2002; 296:155-158, ©2002 The American Association for the Advancement of Science.

clearance of ECM components that are produced during the inflammatory response. Current studies are directed at elucidating the cell surface receptors on macrophages that are involved in recognizing HA degradation products and triggering innate immune responses.

Interferon-γ (IFN-γ) and CXCR3 in Lung Fibrosis

Tissue injury and repair consists of two major components, the inflammatory response and the fibroproliferative response (Fig. 4). We have recently investigated potential links between innate immune responses and fibroproliferation.[12] We have been pursuing the concept that innate immune response may have a role in regulating the fibroproliferative response independent of effects on the inflammatory response. The rationale behind this approach is based on the lung pathology observed in the devastating lung disease Idiopathic Pulmonary Fibrosis (IPF). IPF is a disease that is characterized by unremitting fibroproliferation and deposition of extracellular matrix (Fig. 5). IPF is a progressive disease that results in death from respiratory failure in the majority

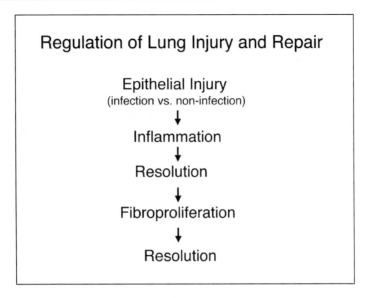

Figure 4. Regulation of lung injury and repair.

of patients within 5 years of diagnosis. The hallmark of this disease is the accumulation of ECM in the lung interstitium with a paucity of inflammation (Fig. 5). The histopathologic pattern of fibrosis suggests that repeated episodes of epithelial cell injury lead to an abnormal wound healing response characterized by foci of myofibroblast accumulation at sites of ECM deposition. We hypothesized that one mechanism that may contribute to an abnormal wound healing response is a failure to properly activate innate immune mechanisms that limit fibroproliferation after lung injury. We suggest that tissue injury initiates two processes simultaneously, one is to activate macrophages to produce key inflammatory mediators required to contain the inciting

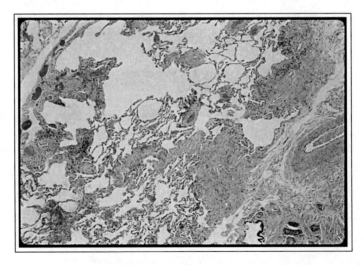

Figure 5. Lung biopsy from a patient with idiopathic pulmonary fibrosis showing the pattern of usual interstitial pneumonia. Reproduced with permission from: Noble PW, Homer RJ. Am J Respir Cell Mol Biol 2005; 33(2):113-120;[13] ©2005 American Thoracic Society.

agent while a second process is simultaneously initiated to limit the extent of fibroproliferation of mesenchymal cells. The goal of the host is to wall off the offending agent, but not destroy organ function in the process. In essence, the fibroproliferative response is designed to replace damaged tissue with matrix components but not destroy the structural integrity of the organ and impair function. In the case of lung injury, this would constitute a limited deposition of collagen. The links between the innate immune system and fibroproliferation have not been extensively investigated. We hypothesized that noninfectious lung injury could activate the immune system in a manner analogous to pathogen-initiated tissue injury. A fundamental component of the immune recognition of invading pathogens and tumor cells is interferon-γ (IFN-γ IFN-γ is produced by a variety of cells including activated T cells and NK cells and is capable of initiating the expression of specific genes in different cell types. IFN-γ has been shown to induce a unique set of chemokines known as CXC chemokines, CXCL9 (Mig), CXCL10 (IP-10) and CXCL11 (ITAC). These chemokines have been shown to have important roles in regulating Th1 responses and have angiostatic and antimicrobial properties. CXCL9, 10 and 11 are recognized exclusively by the chemokine receptor CXCR3. CXCR3 is expressed on activated T cells, dendritic cells and subsets of endothelial cells. We investigated the hypothesis that IFN-γ is produced following noninfectious lung injury and has an important role in regulating fibroproliferation. We further hypothesized that CXCR3 is required for optimal production of IFN-γ Following bleomycin-induced lung injury, there is a brief burst of IFN-γ production that is measurable in the lung of C57BL/6 wild type mice (Fig. 6). This burst of IFN-γ production was severely diminished in CXCR3 null

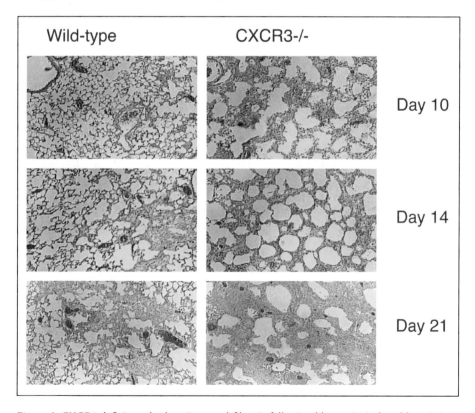

Figure 6. CXCR3 deficiency leads to increased fibrosis following bleomycin-induced lung injury. Reproduced with permission from: Jiang et al, J Clin Invest 2004; 114:291-299, ©2004 The American Society for Clinical Investigation.

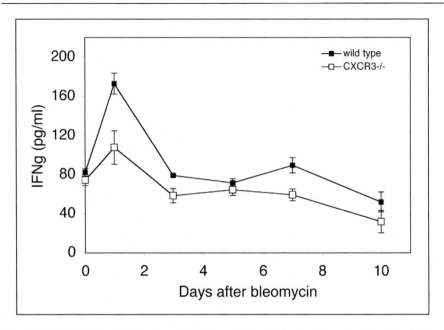

Figure 7. CXCR3 deficiency leads to impaired IFN-γ production following bleomycin-induced lung injury. Reproduced with permission from: Jiang et al, J Clin Invest 2004; 114:291-299, ©2004 The American Society for Clinical Investigation.

mice (Fig. 6). When we examined the injury and repair response in CXCR3 null mice we made the provocative observation that there was substantially increased collagen deposition compared to wild type mice (Fig. 7). Unlike the phenotype observed in the CD44 null mice, the CXCR3 null mice were found to have increased mortality in the absence of an enhanced inflammatory response relative to wild type mice. These mice succumbed to unremitting organ fibrosis. In order to determine if this increased fibrosis was a consequence of the impaired IFN-γ production we took three approaches. First, we administered exogenous IFN-γ only during the first 48 hours to the CXCR3 null mice and found that the fibrotic response returned to wild type levels.[12] In addition we administered IFN-γ neutralizing antibody to the wild type mice and found an enhanced fibrotic response to bleomycin treatment.[12] Finally, the adoptive transfer of CXCR3 expressing immune cells from wild type mice to CXCR3 null mice also reversed the fibrotic phenotype.[12] Collectively, these studies provided evidence that the host innate immune response to noninfectious injury initiated a program of gene expression designed to produce mediators that limit the extent of organ fibrosis.

Summary

We have described two distinct phenotypes that develop in response to a common noninfectious injury to the lung (Fig. 8). CD44 deficiency leads to unremitting lung inflammation and an inability to clear ECM degradation products. The mice succumb to inflammation as a result of a failure to resolve the inflammation and establish limited fibrosis. In contrast, the phenotype observed in the CD44 deficient mice, CXCR3 deficiency resulted in an exaggerated fibrotic response to bleomycin-induced lung injury. The inflammatory response was similar to that found in the wild type mice. This exaggerated fibrotic response in the CXCR3 null mice observed at 21 days was a result of impaired IFN-γ production in the first 48 hours after lung injury. These studies suggest that important insights into tissue responses to injury can be gained by the use of genetically targeted mice and that the innate response of the host may yield important clues to understanding mechanisms of organ fibrosis.

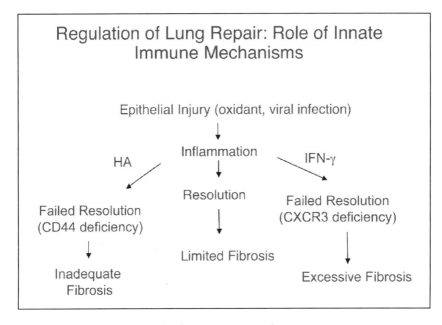

Figure 8. Regulation of lung repair: role of innate immune mechanisms.

Acknowledgements

This work was supported by research grants from the National Institutes of Health (HL-57486 and AI-52478).

References

1. Barton GM, Medzhitov R. Toll-like receptor signaling pathways. Science 2003; 300(5625):1524-1525.
2. Takeda K, Kaisho T, Akira S. Toll-like receptors. Annu Rev Immunol 2003; 21:335-376.
3. Teder P, Vandivier RW, Jiang D et al. Resolution of lung inflammation by CD44. Science 2002; 296(5565):155-158.
4. Fraser JR, Laurent TC, Laurent UB. Hyaluronan: Its nature, distribution, functions and turnover. J Intern Med 1997; 242(1):27-33.
5. Camenisch TD, Spicer AP, Brehm-Gibson T et al. Disruption of hyaluronan synthase-2 abrogates normal cardiac morphogenesis and hyaluronan-mediated transformation of epithelium to mesenchyme. J Clin Invest 2000; 106(3):349-360.
6. Horton MR, Burdick MD, Strieter RM et al. Regulation of hyaluronan-induced chemokine gene expression by IL-10 and IFN-gamma in mouse macrophages. J Immunol 1998; 160(6):3023-3030.
7. McKee CM, Penno MB, Cowman M et al. Hyaluronan (HA) fragments induce chemokine gene expression in alveolar macrophages. The role of HA size and CD44. J Clin Invest 1996; 98(10):2403-2413.
8. Noble PW, McKee CM, Cowman M et al. Hyaluronan fragments activate an NF-kappa B/I-kappa B alpha autoregulatory loop in murine macrophages. J Exp Med 1996; 183(5):2373-2378.
9. Termeer C, Benedix F, Sleeman J et al. Oligosaccharides of Hyaluronan activate dendritic cells via toll-like receptor 4. J Exp Med 2002; 195(1):99-111.
10. Ohkawara Y, Tamura G, Iwasaki T et al. Activation and transforming growth factor-beta production in eosinophils by hyaluronan. Am J Respir Cell Mol Biol 2000; 23(4):444-451.
11. Schmits R, Filmus J, Gerwin N et al. CD44 regulates hematopoietic progenitor distribution, granuloma formation, and tumorigenicity. Blood 1997; 90(6):2217-2233.
12. Jiang D, Liang J, Hodge J et al. Regulation of pulmonary fibrosis by chemokine receptor CXCR3. J Clin Invest 2004; 114(2):291-299.
13. Noble PW, Homer RJ. Back to the future: historical perspective on the pathogenesis of idiopathic pulmonary fibrosis. Am J Respir Cell Mol Biol 2005; 33(2):113-120.

An Eye on Repair:
Myofibroblasts in Corneal Wounds

James V. Jester*

Abstract

Injury to the cornea often leads to corneal fibrosis and scarring resulting in loss of corneal transparency and blindness. Furthermore, current approaches to surgically correct refractive errors, including radial keratotomy (RK), and excimer laser photorefractive keratectomy (PRK) cause damage to the cornea leading to corneal haze and reduced visual acuity. Past studies show that myofibroblasts play a critical role in both the development of wound fibrosis and corneal haze. Myofibroblasts migrate into corneal wounds and establish an interconnected and interwoven contractile network that is necessary for wound contraction, the principal cause of regression following RK. Failure of myofibroblasts to populate the wound results in marked wound gape, mechanical instability and progressive hyperopic refractive changes. Myofibroblasts appearing after PRK also produce marked scattering of light that contributes significantly to corneal haze. In cell culture, myofibroblast differentiate from quiescent stromal cells (keratocytes) after treatment with transforming growth factor-β (TGF-β). Furthermore, treatment of corneal wounds with neutralizing antibodies to TGF-β blocks wound fibrosis and corneal haze after PRK. Recent studies have shown that the differentiation of keratocytes to myofibroblasts involves a synergistic signalling cascade involving integrins, platelet-derived growth factor (PDGF) and TGF-β. These new findings suggest novel therapeutic strategies to modulate myofibroblast differentiation and control corneal wound fibrosis and eliminating corneal haze.

Introduction

Ocular scaring has significant pathophysiological affects causing traction retinal detachment, glaucoma, secondary cataract and corneal opacification. For the lens,[1] retina[2] and cornea[3,4] fibrosis has been associated with the appearance of the 'myofibroblasts' and in the cornea, there is a consensus that corneal myofibroblasts are derived from adjacent corneal stromal cells or 'keratocytes'.[5-7] The transition from quiescent keratocyte to myofibroblast has been shown in the rabbit to be induced by transforming growth factor-β (TGF-β)[5,8] and inhibition of TGF-β using neutralizing antibodies has been shown to block the corneal fibrotic response.[9,10]

Importantly, the cornea as a transparent tissue provides a unique opportunity to study the process of wound healing and corneal myofibroblasts in living tissue at the cellular level using recently developed novel optical imaging approaches combined with digital image analysis. One system that has been used extensively to noninvasively and sequentially study corneal repair in living animals has been in vivo confocal reflectance microscopy (CM). The paradigm of confocal microscopy provides dramatically improved lateral and axial resolution facilitating optical sectioning of living tissue at precise depths and generating quantitative information

*James V. Jester—Department of Ophthalmology, University of California at Irvine, Irvine, California, 92868 U.S.A. Email: JJester@UCI.edu

Tissue Repair, Contraction and the Myofibroblast, edited by Christine Chaponnier, Alexis Desmoulière and Giulio Gabbiani.
©2006 Landes Bioscience and Springer Science+Business Media.

about the axial displacement between different structures. In reflectance based confocal microscopy, images of light scattering structures are obtained such as cells, vessels, nerves and other structures important in characterizing tissue organization.[11-13] In general, in vivo CM has given a more dynamic, "4-dimensional" view of the cornea at a cellular level, which have led to important insights into cellular responses and interactions involved in ocular irritation,[14,15] contact lens wear,[16-18] wound healing,[10,19] and kidney function.[20]

Because of the importance of this relatively new technology, this chapter will begin with a basic review of in vivo CM of the normal cornea. The chapter will then discuss studies of corneal wound healing that have focused on the behaviour and function of the corneal myofibroblast in wound repair following surgically induced injury associated with radial keratotomy (RK) and excimer laser photorefractive keratectomy (PRK).

Corneal Imaging Using in Vivo CM

Minsky in 1957 designed the first confocal microscope that focused illuminating light within a small volume of the specimen using an objective lens and then collected light from this same volume using a similar lens to focus the light back to a detector.[21] Since both the illumination side and the detection side of the microscope were in the same focus, the microscope was referred to as 'confocal'. For a more complete review of the development and design of the in vivo CM the reader is referred to several excellent reviews that cover in more detail the technology and instrumentation that is currently available.[22-24]

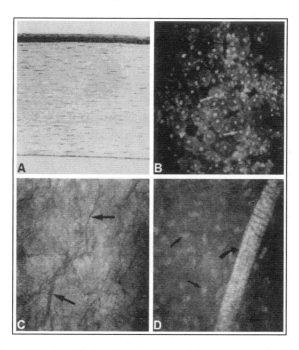

Figure 1. Comparison of the normal light microscopic image (A) to in vivo, confocal reflectance microscopic image of the corneal epithelium (B), basal lamina (C), and stroma (D). Note that images from the confocal microscope provide an en face orientation to the tissue similar to scanning electron microscopic images. In B, surface epithelial cells appear as broad, flat, overlapping cells with brightly reflecting nuclei (arrows). Dark spaces between some cells suggest irregularities in the topography of the surface epithelium. In C, the basal lamina shows a variegated appearance with multiple dark lines (arrows). Below the basal lamina in D, images of the corneal stroma show brightly reflecting keratocyte nuclei (arrows) and occasional nerves (curved arrow). Taken from Jester et al,[26] with permission of Investigative Ophthalmology & Visual Science. Bar = 100 μm.

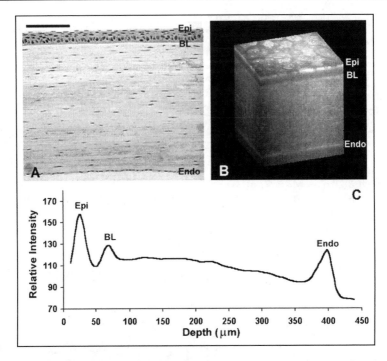

Figure 2. A) Histologic photomicrograph of a normal rabbit cornea showing the corneal epithelium (Epi), basal lamina (BL) and corneal endothelial layer (Endo). B) 3-Dimensional reconstruction of a living rabbit cornea taken from a through focus data set. C) Plot of image pixel intensity versus focal plane depth of the 3-dimensional reconstruction shown in B. Taken from Jester et al,[13] with permission from Elsevier. Bar = 100 μm.

Using in vivo CM the normal intact, living cornea shows distinct light scattering structural features that can be easily detected and correlated to standard normal histologic sections of fixed and processed tissue (Fig. 1A), including the superficial corneal epithelium (Fig. 1B), the epithelial basement membrane (Fig. 1C), and the nuclei of keratocytes and stromal nerves (Fig. 1D). Additionally, images can be collected through the entire living cornea at regular intervals using a confocal microscopy through focusing approach (CMTF); for more details of this technique the reader is referred to original articles by Li et al[25] and Jester et al.[13] The 3-dimensional data set can then be digitally reconstructed to quantitatively measure tissue thickness by generating a cross-sectional (x-z) view of the living cornea that appears similar to that of histologic sections (Fig. 2A,B). An axial, light scattering depth-intensity profile can be generated from the 3-dimensional dataset by calculating the average pixel intensity of each image and plotting image intensity as a function of axial depth (Fig. 2C). This representation identifies the major light scattering structures as peaks in the depth-intensity profile. From these representations, precise measurements of epithelial and stromal thickness and quantitative assessment of corneal light scattering (haze) can be obtained.

Wound Contraction following Incision Corneal Injury

First, corneal wound healing is contrasted from epidermal wound healing by the avascularity of the cornea and the absence of bleeding or formation of a clot. While this makes corneal healing different from healing of skin, the absence of blood allows for detailed imaging of the entire wound during the process of healing. Studies of wound healing show that incisional injury produces an initial gaping wound that is rapidly lined by migrating surface corneal epithelium that fills the wound and may actually enlarge the wound gape by three

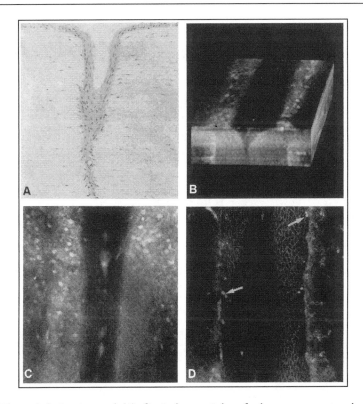

Figure 3. Histopathologic micrograph (A) of incised cornea 3 days after keratotomy compared to confocal microscopic reconstruction of the wound from a living cat eye (B-D). By 3 days, the surface epithelial sheet has migrated into the corneal incision and lines the cut surfaces of the stroma as seen in cross-section using routine histologic processing (A) and the orthogonal cabinet projection (B). In B, single images that comprise the stack can be paged through to reveal details of the cells surrounding the wound (C,D). At the surface of the wound (C) the epithelium appears to be elevated above that overlying the uncut stroma suggesting localized stromal edema. Deeper into the stroma (D) details of the epithelial cell borders within the epithelial plug can be detected suggesting pooling of fluid around the cells and intercellular edema. Between epithelial cells, bright bi-lobed reflections indicate presence of inter-epithelial and stromal inflammatory cells (arrows). (Original magnification: A, x178; B, x190; and C, D, x260). Taken from Jester et al,[26] with permission of Investigative Ophthalmology & Visual Science.

days after injury (Fig. 3).[26] Little or no inflammation is detected following uncomplicated incisional injury and only a few inflammatory cells are seen within the basal epithelium lining the wound margins at this early stage (3D, arrows). Importantly, no fibroblasts are detected within the margins of the wound, and only increased light scattering from keratocytes adjacent to the wound margin can be detected. Over the next week, the incisional gape continues to increase in size as cells adjacent to wound become activated, migrate into the wound bed and replace the epithelium lining the wound margin. By day 14 wounds show the first reduction in wound gape and interestingly also show the final replacement of the epithelial plug with wound healing fibroblasts/myofibroblasts as seen by histopathology and by in vivo CM (Fig. 4A,B, between arrows). Adjacent to the epithelium, wound-healing fibroblasts appear as spindle-shaped cells aligned parallel to the wound margin (Fig. 4C). However, deeper within the wound, fibroblasts are more irregularly arranged and appear to become interwoven; extending between and spanning the wound margins (Fig. 4D, arrow). By 30 days after surgery, there is considerable reduction in wound gape. This reduction

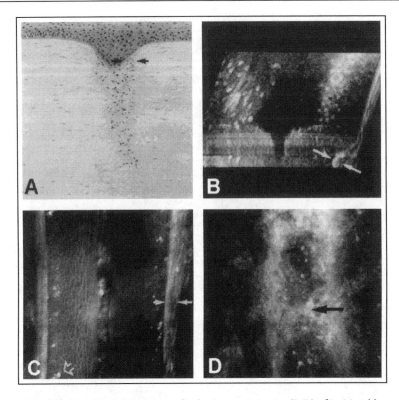

Figure 4. Histopathologic (A) and in vivo, confocal microscopic images (B-D) of incisional keratotomy wounds 14 days after surgery. Confocal images taken of the same wound shown in Figure 5B-D. Initiation of wound contraction correlated with the in-growth of wound healing fibroblasts between the corneal stroma and the epithelial plug (A, between arrows) which was seen by confocal microscopy as a region of reflective cells lining the wound margin (B, between arrows). In individual optical slices (C) the subepithelial area adjacent to the corneal stroma appeared to be composed of spindle shaped clles (between arrows). Within the epithelial plug, persistence of inflammatory cells was detected (open arrow). At the base of the epithelial plug (D) the highly reflective, spindle shaped cells appeared to become more randomly oriented (large arrow) while inflammatory cells or degenerative epithelial cells were detected within the basal epithelial cells (small arrows). (Original magnification: A, x178; B, x190; C, D, x260). Taken from Jester et al,[26] with permission of Investigative Ophthalmology & Visual Science.

correlates with the complete loss of the epithelial plug and replacement with wound healing fibroblasts (Fig. 5A,B) that are highly reflective and difficult to distinguish (Fig. 5C). Deeper within the wound, adjacent keratocytes appeared as an interconnected network that orient toward the wound and interconnect with wound healing fibroblasts (Fig. 5D).

Quantitative measurements of wound gape suggest a biphasic healing response involving a noncontractile and contractile phase depending on the presence, or absence of wound healing fibroblasts with wound gape initially increasing till day 10 and then decreasing from day 14 on. Studies of nonhuman primates, show similar incisional gape changes that correlate with changes in corneal curvature such that increasing wound gape is associated with continued corneal flattening and decreasing wound gape is associated with corneal steeping.[27] Overall, these findings confirm the widely held view that incisional wound gape and peripheral bulging of the cornea is responsible for the flattening effects of RK. However, these important clinical pathologic correlations also point to corneal wound contraction as being largely responsible for the regression of refractive effect seen in patients.

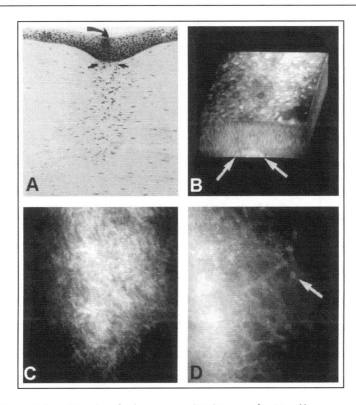

Figure 5. Histopathologic (A) and confocal microscopic (B-D) images of incisional keratotomy wounds 30 days after surgery. Confocal images were taken of the same incision (shown in Figs. 5B-D and 6B-D). Both the histopathologic (A) and the 3-dimensional confocal images (B) show a marked reduction wound gape (between arrows). In the histopathologic micrograph the epithelial surface appears invaginated (A) curved arrow), a similar invagination was not seen by confocal microscopy (B). In single optical sections (C) the area of wound fibrosis appeared as an irregular network of highly reflective structures representing either extracellular matrix or fibroblasts. Deeper within the wound (D) individual keratocyte cell bodies were seen bordering the wound margin (C, arrow). (Original magnification: A, x178; B, x190; C, D, x260). Taken from Jester et al,[26] with permission of Investigative Ophthalmology & Visual Science.

Cellular Mechanism of Wound Contraction in the Cornea

Decreasing wound gape after partial thickness incisional injury was clearly associated with the development of wound fibrosis which involves an initial activation and proliferation of adjacent stromal keratocytes that line the wound margin during the first three days after injury. Fibroblastic cells then migrate into the wound, pushing out the epithelial plug and form a hyper-cellular fibrotic mass within the margins of the wound. Fibroblasts within the wound deposit extracellular matrix, including fibronectin, collagen and other matrix components that then become organized within the wound as the wound contracts (Fig. 6A-C).[3,28,29] Importantly, deposition of extracellular matrix is also associated with a marked increase in the assembly of filamentous actin (Figs. 6D-7F). Detailed confocal microscopic studies of full thickness incisional wounds show that cells migrating into the wound appear as an interconnected meshwork that establishes an interwoven cellular structure throughout the fibrotic tissue (Fig. 7).[30,31] While fibroblasts are initially randomly organized, during wound contraction, fibroblasts become progressively aligned parallel to the wound margins as the wound contracts (Fig. 7D). A similar pattern of random fibroblast organization followed by cell alignment during contraction is also observed following partial thickness incisional injury in the cornea.[30,32]

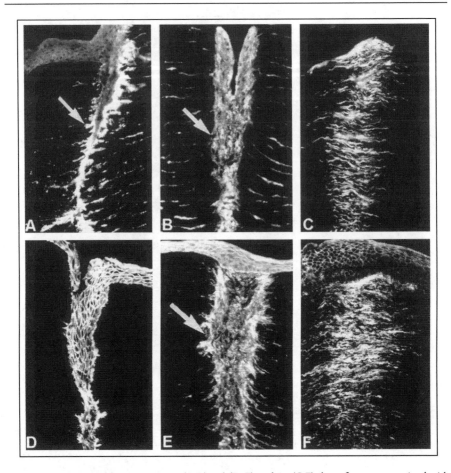

Figure 6. Cat incisional keratotomy at 3 (A,D), 14 (B, E) and 30 (C,F) days after surgery stained with FITC-conjugated goat anti-human fibronectin (A-C) and FITC conjugated phallacidin (D-F). Three days after surgery, anti-fibronectin antibodies (A) stained the stromal wound edge with staining extending along the stromal lamellae (arrow). At the same time, FITC-phallacidin (D) appeared to stain the cortex of the migrating epithelium but only weakly stained adjacent keratocytes. By day 14, developing fibrotic tissue showed intense staining with both anti-fibronectin (B) and phallacidin (E) (arrows). Additionally keratocytes adjacent to the wound showed increased staining but not as intense as that observed in fibroblasts within the wound. At day 30, anti-fibronectin (C) and phallacidin (F) appeared organized into parallel bundles. (Original magnification, x200). Taken from Garana et al,[3] with permission of Investigative Ophthalmology & Visual Science.

Importantly, reorientation of fibroblasts during wound contraction is associated with a change in the organization of contractile actin filament bundles or stress fibers contained within wound healing fibroblasts (Fig. 8).[28,32] Orientation analysis of actin filament assembly in corneal wounds shows that actin filaments are randomly arranged during the early phase of wound healing, similar to the orientation of cells within the wound (Fig. 8A). Bundles of actin filaments detected in one cell, also appear to extend through and connect with actin filament bundles in adjacent cells, perhaps through specific cell-cell adherenes junctional complexes. These findings suggest that cells are interconnected throughout the wound and wound margin forming a syncytium.[30] This possibility is further supported by the finding that functional gap junctions are present between cells within the wound and between cells adjacent to the wound and connecting to the wound

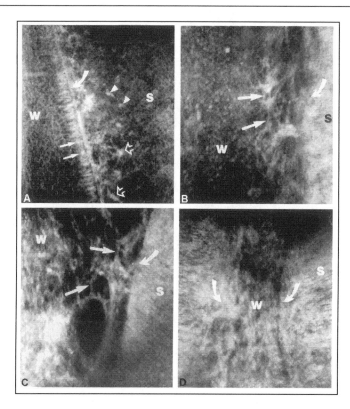

Figure 7. In vivo confocal micrographs taken from the same eye at day 3 (A), 7 (B), 10 (C), and 28 (D) showing the wound area (W), undamaged stroma (S) and the wound margin (curved arrow). At day 3 (A) the wound margin is lined by epithelium. Adjacent to the stroma, basal epithelial cells (arrows) appear highly reflective, while the in the stroma, note the presence of inflammatory cells (arrowheads) and keratocyte nuclei (open arrows) showing early activation and the appearance of the cell body. At day 7 (B) activated keratocytes have migrated into the wound and appear to be interconnected by thin cellular processes (arrows). At day 10 (C), cell process (arrows) appear to be thicker and are aligned predominantly parallel to the wound margin. At day 28 (D) wound contraction has led to near reapposition of the wound margins (curved arrows). (Image width = 400 μm) Taken from Jester et al,[30] with permission from the Royal Microscopical Society.

healing fibroblasts.[30,33] Overall, this appearance further emphasizes the interconnected nature of fibroblasts during the process of wound contraction. As wounds heal and contract, intracellular actin filaments become progressively aligned parallel to the wound margins (Fig. 8D) while at the same time aligning extracellular matrix through mechanical interactions mediated by specific cell-matrix adhesion sites formed to collagen and fibronectin by integral membrane receptors, i.e., integrin α2β1 and α5β1 among others.[3,34,35] Quantitative analysis of these changes also shows a significant correlation between the reorientation of actin filaments from random to aligned with measured changes in wound gape (Fig. 8C,F). This reorientation pattern suggests a unique cellular based contractile mechanism involved in the generation of tension within the wound. Specifically, actin bundles generate force vectors which run parallel to the axis of the bundles and which can be broken down into two components: one oriented across the wound and one oriented parallel to the wound. Since after incisional injury the greatest increase in wound gape occurs across the wound, the least resistance to shortening of wound gape occurs across the wound rather than along the wound. Such a force balance during contraction would tend to twist bundles across the wound as they contract in response to the differences in resistance. This process would

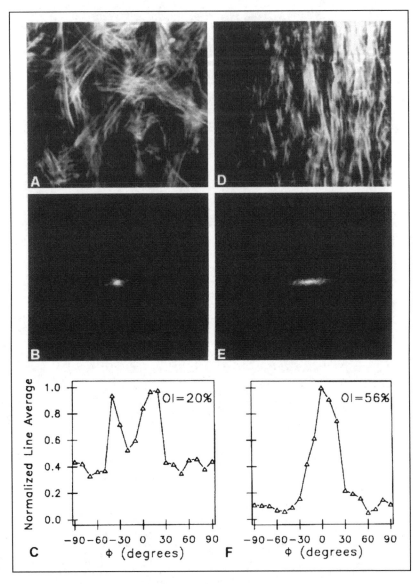

Figure 8. Results of the analysis of microfilament bundle orientation at 10 (A-C) and 28 (D-F) days after injury. A) The 10-day image of a corneal wound appeared to show bundles oriented at several different angles. B) The Fourier Transform (FT) of image A produced a somewhat circular pattern with two angle bands at 20 and -40 degrees which appeared slightly brighter than the rest, most likely corresponding to the parallel groups of stress fibers. C) Line average plot of FT showed two peaks. The orientation index (OI) of 20% demonstrated that there was a slight overall alignment of bundles with the long axis of the wound (an OI of 0% = completely random distribution, 100% = completely parallel to long axis, -100% = completely perpendicular to the long axis). D) The day 28 image of the corneal wound showed alignment of the bundles almost parallel to the long axis. E) The FT produced an elliptical pattern, with the major axis at 90 degrees to the long axis of the wound. F) A plot of the line averages of the FT shows a sharp peak near 0 degrees. The resulting OI of 56% confirmed that microfilament bundles are more oriented along the long axis than the 10-day image. Bar, 25 µm. Taken from Petroll,[32] with permission.

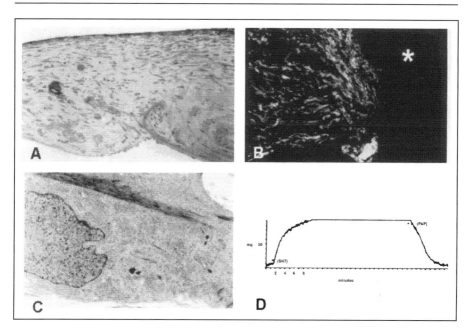

Figure 9. A) Central corneal trephination wound in the rabbit eye, 7 days after injury. The wound contains for the most part a loose connective matrix with fibroblast (Toluidine blue, original magnification x162). B) Frozen section of 7 day old wound stained with NBD-phallacidin. Note the marked staining of cells within the wound and the migration pattern forming pseudo-lamellar layers. Adjacent to the wound within the corneal stroma (asterisk), NBD-phallacidin fluorescence was barely detectable in normal and activated keratocytes. (Original magnification, x250). C) Transmission electron micrograph of corneal wound healing fibroblast 1 month after injury. Note the prominent microfilament bundle containing electron-dense bodies similar to that observed in stress fibers, smooth muscle cells and skin myofibroblasts. (Uranyl acetate and lead citrate, original magnification, x11,000). D) Response of avascular corneal wound tissue to serotonin (5-HT). Note the characteristic contraction response curve and over 100 mg of force generated by the specimen. Relaxation was elicited by addition of papaverine (PAP). Asterisk marks time agents were added. Taken from Luttrull et al, Invest Ophthalmol Vis Sci 1985; 26(10):1449-1452,[40] with permission of Investigative Ophthalmology & Visual Science, and from Jester et al, Am J Pathol 1987; 127(1):140-148,[4] with permission from the American Society for Investigative Pathology.

result in closure of the wound, as well as a more parallel orientation of the bundles along the long axis of the wound, comparable to the tightening of a shoelace.

This model of wound contraction implies an active, intracellular muscle-like contractile mechanism. Gabbiani et al originally described such a contractile mechanism for dermal wound healing myofibroblasts, which have prominent microfilament bundles containing electron dense structures similar to the ultrastructural organization of smooth muscle cells.[36,37] Dermal myofibroblasts also show nuclear indentations similar to contracted smooth muscle cells and exhibit smooth muscle like contractile responses to smooth muscle agonists and express a smooth muscle specific α-isoform of actin (α-SMA).[38,39] Studies of corneal wound healing fibroblasts have identified similar features to the dermal myofibroblast in that cells have prominent microfilament bundles with electron dense structures and show in vitro contractile responses to smooth muscle agonists (Fig. 9).[4,40] Corneal wound healing myofibroblasts also uniquely express smooth muscle specific α-SMA and show a temporal appearance based on expression of α-SMA protein that coincides with the wound contractile response, i.e., maximally expressed at the beginning of wound contraction (day 14) and disappearing by day 28 at the end of wound contraction (Fig. 10).[6]

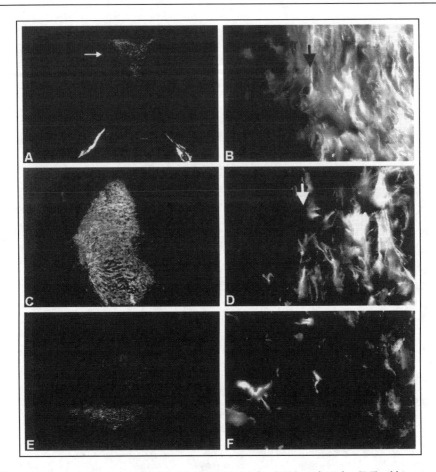

Figure 10. Immunofluorescent staining of 7-day (A,B), 14-day (C,D), and 28-day (E,F) rabbit corneal full-thickness wound by monoclonal antibodies specific for human α-SMA. A) Cells within the anterior part of the wound (arrow) show intense staining. B) *En face* section from the anterior of the wound (arrow indicates wound margin). Only cells within the wound appeared to stain. C) Cross-section of 14-day wound showing that the entire wound is positive for α-SMA. D) *En face* sections showed bundles of α-SMA staining, restricted to the wound (arrow indicates wound margin). E) Cross-section of 28-day wound showed staining primarily in the posterior part of the wound and little staining anteriorly where the wound has already contracted. F) *En face* section from anterior of 28-day wound showed diffuse staining of α-SMA, indicating breakdown of stress fibers in the wound. Also at this time point, cells outside the wound margin appeared to stain for α-SMA indicating possible migration of cells out of the wound. (Original magnification; A, C, E, x52; B, D, F, x720). Taken from Jester et al,[6] with permission of Investigative Ophthalmology & Visual Science.

Taken together, past studies indicate that during the process of wound healing, adjacent corneal keratocytes become activated, proliferate and differentiate to a corneal myofibroblasts. Myofibroblasts then invade the wound as an interconnected and interwoven syncytium that is randomly organized within the wound. During the invasion of the wound, cells synthesize and organize extracellular matrix, collagen type I and III, and fibronectin through integral membrane receptors by exerting force mediated through a smooth muscle-like contractile mechanism. Myofibroblast contraction then both assists in the invasion into the wound and the later contraction of the wound by applying mechanical force to the extracellular matrix and adjacent cells leading to reorientation of cells and matrix parallel to the long axis of the wound margins.

Myofibroblasts, Tissue Growth and Corneal Haze

Excimer surface ablation for PRK removes a substantial amount of corneal stroma for each diopter of refractive change. In rabbits, photoablated tissue is slowly replaced by the deposition of new extracellular matrix between the overlying epithelium and the photoablated stromal bed.[41] In studies using in vivo CM, a similar regrowth of corneal stroma has been detected (Fig. 11). Comparing the 3-dimensional reconstructions of the same cornea taken before and after PRK, it is clear that the marked thinning of the cornea detected at 1 week after surgery (Fig. 11B) is almost completely regrown by 17 weeks after surgery (Fig. 11F). This regrowth of the corneal stroma in the rabbit is associated with several important events that occur during the wound healing response. First, at 1 week after surgery there are two highly light scattering bands in the anterior stroma that are separated by approximately 100 μm of very low light scattering corneal stroma (Fig. 11B, double arrows). The most anterior band is associated with

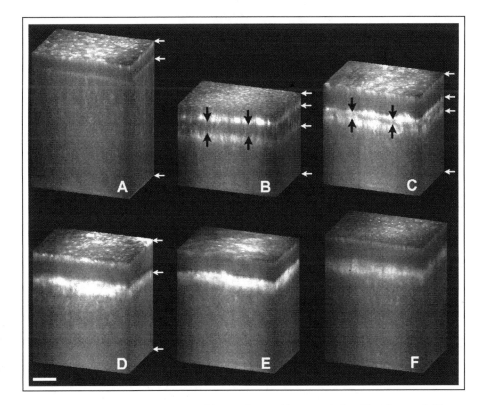

Figure 11. 3-Dimensional reconstructions of the same living rabbit cornea before (A) and one week (B), two weeks (C), 3 weeks (D), 7 weeks (E) and 17 weeks (F) after excimer laser photoablation of the central cornea. In the preoperative cornea (A), three reflective layers were detected corresponding to the superficial epithelium, the epithelial basal lamina, and the endothelium (white arrows). One week (B) and two weeks (C) after injury showed two new reflective layers corresponding anteriorly to the photoablated stromal surface and posteriorly to the migrating stromal fibroblasts (black arrows). Note the dark band between the photoablated surface and migrating cells, which corresponded to the acellular zone in the anterior stroma. Bar = 100 μm in the x-axis and 70 μm in the z-axis. Reproduced from Moller-Pedersen et al,[19] with permission of Investigative Ophthalmology & Visual Science.

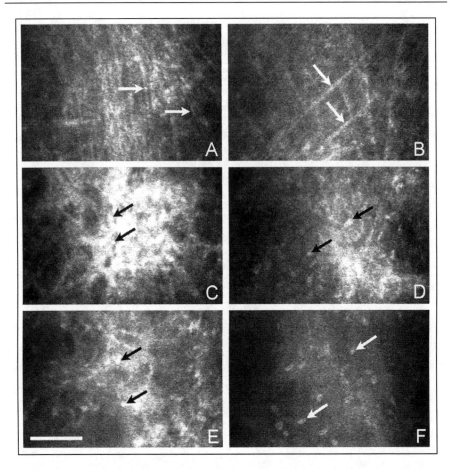

Figure 12. In vivo morphology of vehicle (*left*) and anti-TGF-β (*right*) treated rabbit corneas. Two weeks post-PRK: the density of repopulating, spindle-shaped keratocytes (arrows) appeared higher in vehicle treated corneas (A) than in anti-TGF-β treated corneas (B). Three weeks post-PRK: the anterior stroma of vehicle treated corneas (C) had a high density of reflective, stellate cells with rounded, highly reflective nuclei (arrows), suggesting myofibroblast transformation, compared to the more normal appearing keratocytes (arrows) in anti-TGF-β treated corneas (D). Two months post-PRK: cellularity (arrows) and reflectivity of the anterior stroma had decreased in both groups; however, vehicle treated corneas (E) still appeared more disorganised, hyper-cellular, and reflective compared to anti-TGF-β treated corneas (F). Bar indicates 100 μm. Taken from Moller-Pedersen et al,[10] with permission.

light scattering from the photoablated stromal surface of the cornea. The second light scattering band is associated with the appearance of spindle shaped structures that have been identified as migrating keratocytes (Fig. 12A, arrows).[10,42,43] The region separating these two bands is an acellular zone caused by keratocyte apoptosis or necrosis immediately after surgery.[44,45] It is apparent from this data that keratocytes adjacent to the zone of injury become activated, elongate into spindle shaped cells and then migrate into and repopulate over the first three weeks the region of keratocyte apoptosis. This repopulation can be better appreciated in the 3-dimensional reconstructions as the lower band becomes progressively closer to the photoablated stromal surface at 2 week (Fig. 11C, double arrows) and then merges with the stromal surface by three weeks (Fig. 11D, double arrows).

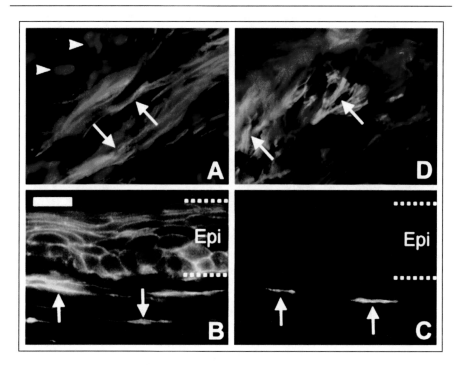

Figure 13. Immunofluorescent staining of rabbit corneas one (A) and two-weeks (B-D) post-PRK showing expression of f-actin (stress fibers) and α-SM-actin (marker for myofibroblast transformation) within wound healing keratocytes. A) Pseudocolored image of en face section showing colocalization of f-actin (green) and keratocyte nuclei (red; propidium iodide counter-staining). Note that the keratocytes appear extremely elongated and show f-actin organised into prominent microfilament bundles, i.e., stress fibers (arrows). Finally, note that adjacent cells show a weak, cortical f-actin organization (arrowheads) suggestive of quiescent keratocytes. B) Cross-section showing intense expression of f-actin in anterior wound healing fibroblasts (arrows) as well as in the overlying epithelium (between dotted lines). C) Same section as B showing colocalization of α-SM-actin (arrows). Only wound healing keratocytes within the photoablation zone appear to stain for both f-actin and α-SM-actin indicating myofibroblast transformation, while keratocytes below (arrowheads) did not stain for α-SM-actin (compare B and C). D) Pseudocolored image of en face section showing colocalization of f-actin (red) and α-SM actin (green). Note that the cells stain intensely for α-SM actin (arrows) and have a more stellate cell morphology analogous to the in vivo findings. Bar indicates 25 µm. Taken from Moller-Pedersen et al,[10] with permission.

When cellular repopulation is complete, migrating keratocytes appear to enlarge, show broadened pseudopodia, and become highly reflective (Fig. 12C, arrows). These changes are morphologically consistent with the differentiation of migrating keratocytes to a myofibroblast phenotype and, importantly, cells at the interface of the wound express α-SMA, the phenotypic marker for myofibroblast differentiation (Fig. 13).[10,45] Brightly reflecting myofibroblasts following PRK appear to persist in the wound for the first month after surgery, showing a progressive decrease in light scattering over time (Figs. 11D-F; 12E). Interestingly, the appearance of myofibroblasts temporally correlates with a marked increase in the amount of stromal haze, which peaks in the rabbit from 2-3 weeks after surgery.[10,46] Furthermore, the disappearance of myofibroblasts also correlates with reduced haze. Overall, the reflectivity of the myofibroblast, appearing high at 2-3 weeks and subsiding by 2 to 3 months, temporally correlates with the previously observed haze.

The appearance of myofibroblasts at the interface of the wound is also associated with beginning growth of the corneal stroma. In comparing 3-dimensional reconstruction from 1, 2, 3 and 7 weeks (Fig. 11B-E) it can be noted that the stromal thickness does not noticeably change until the appearance of myofibroblast at 3 weeks. Myofibroblasts are therefore not only associated with increasing corneal haze but also the initiation of stromal growth.

TGF-β and Appearance of Myofibroblasts in Corneal Wounds

As will be discussed below, TGF-β is known to be a critical regulator of myofibroblast differentiation. When neutralizing antibodies to TGF-β are applied topically to rabbit eyes after PRK there is a marked decrease in the apparent number of spindle shaped, migratory keratocytes within the wound compared to untreated control eyes within the first week after surgery (Fig. 12B, arrows).[10] More importantly, after repopulation of the acellular stroma there are fewer cells that appeared to become hyper-reflective (Fig. 12D, arrows) and those that did showed a more rapid return or were replaced by a more normal keratocyte population with markedly reduced light scattering (Fig. 12F, arrows). Furthermore, corneal "haze" following treatment was significantly reduced both in extent (peak haze) and duration (time to return to baseline) (Fig. 14).

Overall, these observations support the conclusion that TGF-β plays an important role in controlling myofibroblast transformation following PRK and that blocking antibodies to TGF-β can inhibit this differentiation process. Furthermore, TGF-β mediated myofibroblast

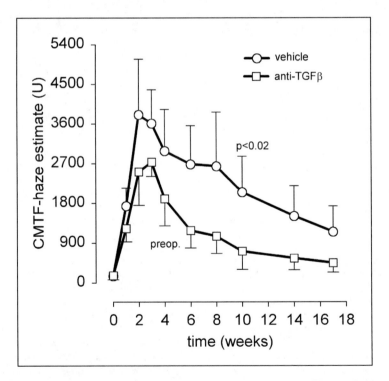

Figure 14. Comparison of stromal light reflectivity in vehicle (°) and anti-TGF-β (□) treated rabbit corneas following PRK. Anti-TGF-β treated corneas developed about 30% less intense reflections during the first three weeks and the intensity declined at a higher rate, leading to about 60% reduction during the rest of the study (ANOVA, p<0.02). CMTF-haze estimate is expressed in arbitrary units (U) defined as μm * pixel intensity. Data are mean ± SD, n = 5. Taken from Moller-Pedersen et al[10] with permission from Oxford University Press.

differentiation appears to play an important role in the development of corneal "haze" after PRK since blocking myofibroblast differentiation significantly reduces corneal haze. Our understanding of why myofibroblasts show greatly increased light scattering remains unclear. While the popular explanation for corneal haze remains focused on the extracellular matrix, recent studies suggest that the property of transparency is not limited to the extracellular matrix but also involves critical biophysical characteristics of the corneal keratocytes themselves. Recent studies have shown that corneal cells, epithelial and keratocytes, abundantly express a few water-soluble proteins to levels similar to that observed for the enzyme crystallin proteins of the lens.[47,48] Some investigators have proposed that these proteins at high concentrations may have similar biophysical properties to lens crystallin proteins and may contribute to the transparent state of the cornea by destructively interfering with scattered light through short-range physical interactions.[48,49] In support of this novel theory, water-soluble proteins isolated from rabbit keratocytes obtained from transparent rabbit corneas show high levels of expression of two corneal crystallins, aldehyde dehydrogenase class 1 (ALDH1A1) and transketolase (TKT). When cells are obtained from hazy corneas containing light scattering fibroblasts and myofibroblasts the level of expression of ALDH1A1 and TKT is markedly reduced or abolished compared to the level of expression within keratocytes from transparent regions immediately adjacent to the hazy corneal regions. These findings point to the possibility that corneal haze following refractive surgery represents a defect in the light scattering properties of myofibroblasts and is not related to collagen deposition or fibrosis. This possibility is critically important given that transparency may be easily reestablished in cells but permanently lost with corneal fibrosis.

Differentiation of Keratocytes to Myofibroblasts

Considerable progress has been made in identifying the molecular mechanisms underlying differentiation of corneal keratocytes to myofibroblasts. Earlier studies in skin indicate TGF-β plays an important role in up regulating matrix synthesis and the appearance of myofibroblasts in dermal wounds and the expression of α-SMA in cultured quiescent dermal fibroblasts.[50] Additionally, application of neutralizing antibodies to TGF-β isoforms effectively block dermal wound fibrosis.[51-53] Likewise, studies of corneal keratocytes show similar up-regulation of α-SMA after addition of TGF-β to the culture media (Fig. 15A).[5] Sparsely cultured corneal keratocytes also differentiate to myofibroblasts without exogenous TGF-β,[7] however this effect is most likely mediated by autocrine expression of TGF-β when cells lose their cell-cell attachments.[54] In a culture model of corneal keratocyte injury, scrape wounds induced expression of TGF-β at the wound edge, suggesting that incisional injuries induce expression of TGF-β by cells immediately adjacent to the wound margin, leading to TGF-β mediated myofibroblast differentiation.

However, the cellular and molecular mechanisms of myofibroblast differentiation remain somewhat unclear. In wounds, myofibroblasts appear only within the margins of the wound and are not detected outside, suggesting important cell-matrix interactions involved in regulating myofibroblast differentiation.[6] Recently, Tomasek has proposed that mechanical tension within the wound plays a critical role in directing protomyofibroblast and myofibroblast differentiation within tissues.[55] In support of this theory is the finding that inhibiting cell-matrix interactions by interfering with integrin-fibronectin binding using small peptide blocking sequences (RGD), inhibits myofibroblast differentiation of corneal keratocytes.[56] Blocking of downstream signalling through integrin receptor ligation using tyrosine kinase inhibitors also blocks TGF-β induced myofibroblast differentiation. More recent studies support these earlier findings and indicate that focal adhesion kinase activity (FAK) is critical to the downstream myofibroblast differentiation and regulation of α-SMA expression.[57] Together, these observations suggest that outside-in signalling through integrin receptors that are responding to important mechanical signals are critical to myofibroblast differentiation.

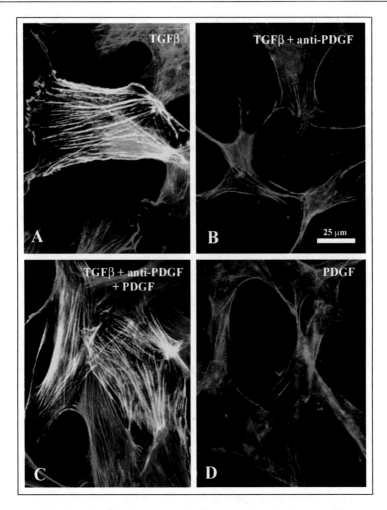

Figure 15. Colocalization of α-SM actin (green) with filamentous actin (red) in corneal keratocytes treated with A. TGF-β (1 ng/ml), B. TGF-β +anti-PDGF (25 ug/ml), C. TGF-β + anti-PDGF + PDGF (100 ng/ml) and D. PDGF alone. Note the absence of α-SM actin within keratocytes treated with TGF-β in combination with anti-PDGF and PDGF treated alone. Taken from Jester et al,[60] with permission from Elsevier.

Besides cell-matrix interactions, other growth factors appear to play an important role in myofibroblast differentiation. Previous studies of TGF-β induced proliferative responses have shown that cell division is regulated by a TGF-β induced PDGF autocrine feedback loop.[58,59] In exploring the effects of PDGF on myofibroblast differentiation, neutralizing antibodies to PDGF blocked not only cell proliferation, but also expression of α-SMA and myofibroblast differentiation as indicated by the failure to spread and organize intracellular actin filament bundles both characteristic of myofibroblast differentiation in culture (Fig. 15B).[60] While addition of exogenous PDGF to antibody treated cultures overcomes this inhibition (Fig. 15C), PDGF alone does not induce expression of α-SMA nor markedly increase actin filament assembly and cell spreading (Fig. 15D). Rather, PDGF treatment of culture keratocytes induces cell elongation to form spindle shaped fibroblastic cells that are markedly different than the enlarged, spread myofibroblasts. Together, the observations support a synergistic

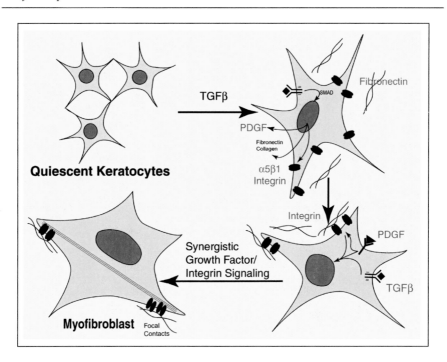

Figure 16. Molecular pathway for TGF-β induction of myofibroblast differentiation. Taken from Jester et al,[60] with permission from Elsevier.

interaction between extracellular matrix and matrix receptors, TGF-β and TGF-β receptors and an autocrine PDGF feedback loop (Fig. 16). Based on this model injury up-regulates expression of TGF-β by keratocytes adjacent to the wound margin. Alternatively, recent studies have shown that TGF-β2 is expressed by the corneal epithelium overlying regions of stromal injury, suggesting that the corneal epithelium may be an important source of TGF-β for the induction of myofibroblast differentiation.[61] This is consistent with the finding that epithelial migration along the wound margin and plug formation appear to be an important first step in incisional wound healing and induction of wound fibrosis. TGF-β acting on adjacent keratocytes then induces the expression of α5β1 integrin, fibronectin and PDGF through the classical SMAD signal transduction pathway.[62] Expression of these extracellular matrix proteins, receptors and growth factors then establish a synergistic growth factor signalling cascade that is perhaps modulated by critical mechanical environmental conditions characterizing the wound. While the exact signal transduction cascade has yet to be established, keratocytes become activated, proliferate, spread and assemble a contractile actin filament network that simultaneously induces expression of α-SMA, characteristic of myofibroblast differentiation.

There is still much to learned concerning the molecular mechanisms controlling wound repair and myofibroblast differentiation. Other growth factors clearly play important roles in this process, and their mechanisms of action have yet to be clearly identified. In particular, recent studies suggest that connective tissue growth factor (CTGF), which is upregulated by TGF-β and is expressed by corneal fibroblasts in vitro and during healing,[63-65] plays an important role in modulating myofibroblast contractile activity.[66] Additionally, hepatocyte growth factor and keratinocyte growth appear to play important roles in mediating epithelial-keratocyte interaction differentially regulating growth and differentiation.[67,68] How these growth factors modulate the wound healing response is unclear and future studies are needed.

Acknowledgements

The author wants to recognize the contributions of his long time collaborators, W. Matthew Petroll and H. Dwight Cavanagh, that together have lead to the many exciting findings and great times spent discussing myofibroblast differentiation and function in the cornea. Additionally, this work was support in part by grants for the National Eye Institute EY07348 and a Senior Scientist Award and Challenge Grant from Research to Prevent Blindness Inc, New York, New York.

References

1. Novotyn GE, Pau H. Myofibrolbast-like cells in human anterior capsular cataract. Virchows Archiv A 1984; 404:393-401.
2. Kampik A, Kenyon KR, Michels RG et al. Epitretinal and vitreous membranes. Comparative study of 56 cases. Arch Ophthalmol 1981; 99:1445-1454.
3. Garana RM, Petroll WM, Chen WT et al. Radial keratotomy II. Role of the myofibroblast in corneal wound contraction. Invest Ophthal Vis Sci 1992; 33(12):3271-3282.
4. Jester JV, Rodrigues MM, Herman IM. Characterization of avascular corneal wound healing fibroblasts. New insights into the myofibroblast. Am J Pathol 1987; 127(1):140-148.
5. Jester JV, Barry-Lane PA, Cavanagh HD et al. Induction of alpha-smooth muscle actin expression and myofibroblast transformation in cultured corneal keratocytes. Cornea 1996; 15(5):505-516.
6. Jester JV, Petroll WM, Barry PA et al. Expression of alpha-smooth muscle (alpha-SM) actin during corneal stromal wound healing. Invest Ophthal Vis Sci 1995; 36(5):809-819.
7. Masur SK, Dewal HS, Dinh TT et al. Myofibroblasts differentiate from fibroblasts when plated at low density. Proc Natl Acad Sci USA 1996; 93(9):4219-4223.
8. Petridou S, Masur SK. Immunodetection of connexins and cadherins in corneal fibroblasts and myofibroblasts. Invest Ophthalmol Vis Sci 1996; 37(9):1740-1748.
9. Jester JV, Barry-Lane PA, Petroll WM et al. Inhibition of corneal fibrosis by topical application of blocking antibodies to TGF beta in the rabbit. Cornea 1997; 16(2):177-187.
10. Moller-Pedersen T, Cavanagh HD, Petroll WM et al. Neutralizing antibody to TGFbeta modulates stromal fibrosis but not regression of photoablative effect following PRK. Curr Eye Res 1998; 17(7):736-747.
11. Cavanagh HD, Petroll WM, Jester JV. The application of confocal microscopy to the study of living systems. Neurosci Biobehav Rev 1993; 17(4):483-498.
12. Jester JV, Andrews PM, Petroll WM et al. In vivo, real-time confocal imaging. J Electron Microsc Tech 1991; 18(1):50-60.
13. Jester JV, Petroll WM, Cavanagh HD. Measurement of tissue thickness using confocal microscopy. Methods Enzymol 1999; 307:230-245.
14. Jester JV, Petroll WM, Bean J et al. Area and depth of surfactant-induced corneal injury predicts extent of subsequent ocular responses. Invest Ophthalmol Vis Sci 1998; 39(13):2610-2625.
15. Jester JV, Li HF, Petroll WM et al. Area and depth of surfactant-induced corneal injury correlates with cell death. Invest Ophthalmol Vis Sci 1998; 39(6):922-936.
16. Cavanagh HD, Ladage PM, Li SL et al. Effects of daily and overnight wear of a novel hyper oxygen-transmissible soft contact lens on bacterial binding and corneal epithelium: A 13-month clinical trial. Ophthalmology 2002; 109(11):1957-1969.
17. Imayasu M, Petroll WM, Jester JV et al. The relation between contact lens oxygen transmissibility and binding of Pseudomonas aeruginosa to the cornea after overnight wear. Ophthalmology 1994; 101(2):371-388.
18. Ladage PM, Yamamoto K, Ren DH et al. Effects of rigid and soft contact lens daily wear on corneal epithelium, tear lactate dehydrogenase, and bacterial binding to exfoliated epithelial cells. Ophthalmology 2001; 108(7):1279-1288.
19. Moller-Pedersen T, Li HF, Petroll WM et al. Confocal microscopic characterization of wound repair after photorefractive keratectomy. Invest Ophthalmol Vis Sci 1998; 39(3):487-501.
20. Andrews PM, Petroll WM, Cavanagh HD et al. Tandem scanning confocal microscopy (TSCM) of normal and ischemic living kidneys. Am J Anat 1991; 191(1):95-102.
21. Minsky M. Memoir on inventing the confocal scanning microscope. Scanning J 1988; 10:128-138.
22. Cavanagh HD, El-Agha MS, Petroll WM et al. Specular microscopy, confocal microscopy, and ultrasound biomicroscopy: Diagnostic tools of the past quarter century. Cornea 2000; 19(5):712-722.
23. Bohnke M, Masters BR. Confocal microscopy of the cornea. Prog Retin Eye Res 1999; 18(5):553-628.

24. Petroll WM, Cavanagh HD, Jester JV. Clinical confocal microscopy. Curr Opin Ophthalmol 1998; 9(4):59-65.
25. Li HF, Petroll WM, Moller-Pedersen T et al. Epithelial and corneal thickness measurements by in vivo confocal microscopy through focusing (CMTF). Curr Eye Res 1997; 16(3):214-221.
26. Jester JV, Petroll WM, Feng W et al. Radial keratotomy. 1. The wound healing process and measurement of incisional gape in two animal models using in vivo confocal microscopy. Invest Ophthalmol Vis Sci 1992; 33(12):3255-3270.
27. Petroll WM, New K, Sachdev M et al. Radial keratotomy. III. Relationship between wound gape and corneal curvature in primate eyes. Invest Ophthalmol Vis Sci 1992; 33(12):3283-3291.
28. Petroll WM, Cavanagh HD, Jester JV. Assessment of stress fiber orientation during healing of radial keratotomy wounds using confocal microscopy. Scanning 1998; 20(2):74-82.
29. Ljubimov AV, Alba SA, Burgeson RE et al. Extracellular matrix changes in human corneas after radial keratotomy. Exp Eye Res 1998; 67(3):265-272.
30. Jester JV, Petroll WM, Barry PA et al. Temporal, 3-dimensional, cellular anatomy of corneal wound tissue. J Anat 1995; 186(Pt 2):301-311.
31. Petroll WM, Cavanagh HD, Jester JV. Three-dimensional imaging of corneal cells using in vivo confocal microscopy. J Microsc 1993; 170(Pt 3):213-219.
32. Petroll WM, Cavanagh HD, Barry P et al. Quantitative analysis of stress fiber orientation during corneal wound contraction. J Cell Sci 1993; 104(Pt 2):353-363.
33. Watsky MA. Keratocyte gap junctional communication in normal and wounded rabbit corneas and human corneas. Invest Ophthalmol Vis Sci 1995; 36(13):2568-2576.
34. Masur SK, Conors Jr RJ, Cheung JK et al. Matrix adhesion characteristics of corneal myofibroblasts. Invest Ophthalmol Vis Sci 1999; 40(5):904-910.
35. Masur SK, Cheung JK, Antohi S. Identification of integrins in cultured corneal fibroblasts and in isolated keratocytes. Invest Ophthalmol Vis Sci 1993; 34(9):2690-2698.
36. Gabbiani G, Hirschel BJ, Ryan GB et al. Granulation tissue as a contractile organ. A study of structure and function. J Exp Med 1972; 135:719-734.
37. Gabbiani G, Ryan GB, Majno G. Presence of modified fibroblasts in granulation tissue and their possible role in wound contraction. Experientia 1971; 27:549-550.
38. Darby I, Skalli O, Gabbiani G. Alpha-smooth muscle actin is transiently expressed by myofibroblasts during experimental wound healing. Lab Invest 1990; 63:21-29.
39. Gabbiani G. Modulation of fibroblastic cytoskeletal features during wound healing and fibrosis. Pathol Res Pract 1994; 190(9-10):851-853.
40. Luttrull JK, Smith RE, Jester JV. In vitro contractility of avascular corneal wounds in rabbit eyes. Invest Ophthalmol Vis Sci 1985; 26(10):1449-1452.
41. Tuft SJ, Zabel RW, Marshall J. Corneal repair following keratectomy. A comparison between conventional surgery and laser photoablation. Invest Ophthalmol Vis Sci 1989; 30(8):1769-1777.
42. Ichijima H, Petroll WM, Jester JV et al. In vivo confocal microscopic studies of endothelial wound healing in rabbit cornea. Cornea 1993; 12(5):369-378.
43. Chew SJ, Beuerman RW, Kaufman HE. Real-time confocal microscopy of keratocyte activity in wound healing after cryoablation in rabbit corneas. Scanning 1994; 16(5):269-274.
44. Helena MC, Baerveldt F, Kim WJ et al. Keratocyte apoptosis after corneal surgery. Invest Ophthalmol Vis Sci 1998; 39(2):276-283.
45. Mohan RR, Hutcheon AE, Choi R et al. Apoptosis, necrosis, proliferation, and myofibroblast generation in the stroma following LASIK and PRK. Exp Eye Res 2003; 76(1):71-87.
46. Moller-Pedersen T, Cavanagh HD, Petroll WM et al. Corneal haze development after PRK is regulated by volume of stromal tissue removal. Cornea 1998; 17(6):627-639.
47. Sax CM, Salamon C, Kays WT et al. Transketolase is a major protein in the mouse cornea. J Biol Chem 1996; 271(52):33568-33574.
48. Jester JV, Moller-Pedersen T, Huang J et al. The cellular basis of corneal transparency: Evidence for 'corneal crystallins'. J Cell Sci 1999; 112(Pt 5):613-622.
49. Piatigorsky J. Gene sharing in lens and cornea: Facts and implications. Prog Retin Eye Res 1998; 17(2):145-174.
50. Desmouliere A, Geinoz A, Gabbiani F et al. Transforming growth factor-beta 1 induces alpha-smooth muscle actin expression in granulation tissue myofibroblasts and in quiescent and growing cultured fibroblasts. J Cell Biol 1993; 122(1):103-111.
51. Shah M, Foreman DM, Ferguson MW. Neutralisation of TGF-beta 1 and TGF-beta 2 or exogenous addition of TGF-beta 3 to cutaneous rat wounds reduces scarring. J Cell Sci 1995; 108(Pt 3):985-1002.

52. Shah M, Foreman DM, Ferguson MW. Control of scarring in adult wounds by neutralising antibody to transforming growth factor beta. Lancet 1992; 339(8787):213-214.

53. Shah M, Foreman DM, Ferguson MW. Neutralising antibody to TGF-beta 1,2 reduces cutaneous scarring in adult rodents. J Cell Sci 1994; 107(Pt 5):1137-1157.

54. Song QH, Singh RP, Richardson TP et al. Transforming growth factor-beta1 expression in cultured corneal fibroblasts in response to injury. J Cell Biochem 2000; 77(2):186-199.

55. Tomasek JJ, Gabbiani G, Hinz B et al. Myofibroblasts and mechano-regulation of connective tissue remodelling. Nat Rev Mol Cell Biol 2002; 3(5):349-363.

56. Jester JV, Huang J, Barry-Lane PA et al. Transforming growth factor(beta)-mediated corneal myofibroblast differentiation requires actin and fibronectin assembly. Invest Ophthalmol Vis Sci 1999; 40(9):1959-1967.

57. Thannickal VJ, Lee DY, White ES et al. Myofibroblast differentiation by transforming growth factor-beta1 is dependent on cell adhesion and integrin signaling via focal adhesion kinase. J Biol Chem 2003; 278(14):12384-12389.

58. Soma Y, Grotendorst GR. TGF-B stimulates primary human skin fibroblast DNA synthesis via an autocrine production of PDGF-related peptides. J Cell Physiol 1989; 140:246-253.

59. Leof EB, Proper JA, Goustin AS et al. Induction of c-sis mRNA and activity similar to platelet-derived growth factor by transforming growth factor beta: A proposed model for indirect mitogenesis invovling autocrine activity. PNAS 1986; 83:2453-2457.

60. Jester JV, Huang J, Petroll WM et al. TGFbeta induced myofibroblast differentiation of rabbit keratocytes requires synergistic TGFbeta, PDGF and integrin signaling. Exp Eye Res 2002; 75(6):645-657.

61. Stramer BM, Zieske JD, Jung JC et al. Molecular mechanisms controlling the fibrotic repair phenotype in cornea: Implications for surgical outcomes. Invest Ophthalmol Vis Sci 2003; 44(10):4237-4246.

62. Evans RA, Tian YC, Steadman R et al. TGF-beta1-mediated fibroblast-myofibroblast terminal differentiation-the role of Smad proteins. Exp Cell Res 2003; 282(2):90-100.

63. Ivarsen A, Laurberg T, Moller-Pedersen T. Characterisation of corneal fibrotic wound repair at the LASIK flap margin (see comment). Br J Ophthalmol 2003; 87(10):1272-1278.

64. Folger PA, Zekaria D, Grotendorst G et al. Transforming growth factor-beta-stimulated connective tissue growth factor expression during corneal myofibroblast differentiation. Invest Ophthalmol Vis Sci 2001; 42(11):2534-2541.

65. Blalock TD, Duncan MR, Varela JC et al. Connective tissue growth factor expression and action in human corneal fibroblast cultures and rat corneas after photorefractive keratectomy. Invest Ophthalmol Vis Sci 2003; 44(5):1879-1887.

66. Garrett Q, Khaw PT, Blalock TD et al. Involvement of CTGF in TGF-beta1-stimulation of myofibroblast differentiation and collagen matrix contraction in the presence of mechanical stress. Invest Ophthalmol Vis Sci 2004; 45(4):1109-1116.

67. Lee JS, Liu JJ, Hong JW et al. Differential expression analysis by gene array of cell cycle modulators in human corneal epithelial cells stimulated with epidermal growth factor (EGF), hepatocyte growth factor (HGF), or keratinocyte growth factor (KGF). Curr Eye Res 2001; 23(1):69-76.

68. Wilson SE, Weng J, Chwang EL et al. Hepatocyte growth factor (HGF), keratinocyte growth factor (KGF), and their receptors in human breast cells and tissues: alternative receptors. Cell Mol Biol Res 1994; 40(4):337-350.

69. Jester JV, Petroll WM, Cavanagh HD. Corneal stromal wound healing in refractive surgery: The role of myofibroblasts. Prog Retin Eye Res 1999; 18(3):311-356.

Index

A

α smooth muscle actin (α-SMA) 1-4, 16, 17, 25-27, 30, 32, 35, 36, 41-43, 45, 51, 54, 57, 59, 61, 62, 68, 70, 75-78, 82, 89-91, 95, 103, 104, 106, 107, 127, 128, 131, 133-135
Actin isoform 1, 2
Activated stellate cell 92
Adhesion molecule 111
Alignment 10, 11, 17-20, 123, 125, 126
Alport syndrome 47
Alveolar type II cell 68, 69
Angiogenesis 53, 54, 56, 60, 62, 64, 95, 97
Anisotropy 7, 10, 16-18, 20
Apoptosis 2, 4, 49, 57, 77, 82, 94, 102-107, 130
Asthma 1, 40-43, 45, 111

B

Basement membrane (BM) 40, 42-44, 47, 48, 50, 51, 55, 106, 120
Bleomycin 26, 68-71, 111-113, 115, 116

C

Cancer 4, 5, 74-83, 88-97, 99
CD40 35, 37
CD44 110-113, 116
Cell orientation 20, 124, 129
Collagen 1, 2, 7, 9-22, 25-30, 33-35, 40-43, 45, 47-49, 51, 55, 58, 59, 68, 70, 71, 75, 77, 79-82, 92, 102, 103, 106, 107, 111, 115, 116, 123, 125, 128, 133
    Collagen gel 11, 18-20, 27-30, 77, 79
    Collagen remodelling 18, 20
    Collagen type I 2, 80, 82, 92, 128
Colon cancer 75-79, 81-83, 89, 97
Confocal microscopy 118-120, 122, 123
Connective tissue growth factor (CTGF) 2, 25-30, 37, 57, 58, 135
Connective tissue remodeling 1, 4, 20
Contraction 1-5, 10-16, 20, 25, 27-30, 88, 89, 118, 120, 122-125, 127, 128
Cornea 10, 33, 118-123, 129-133, 136
Cross-signaling 74, 75, 82, 83
CXCR3 110, 113, 115, 116

Cyclooxygenase (COX) 34, 35, 93, 97, 98
    COX-2 34, 35, 93, 97, 98
Cytomechanics 7, 8, 9, 11, 13, 16-18, 20, 22

D

Differentiation 1-3, 16, 25, 26, 33, 35, 40-43, 45, 68-72, 118, 131-136

E

Epithelial-mesenchymal transition (EMT) 45, 48, 51, 53-55, 57-59, 61-64, 91
Extracellular matrix (ECM) 2-5, 7-11, 13-17, 20, 22, 25, 26, 30, 32, 33, 40, 42, 45, 48, 54, 55, 61-63, 68-70, 80, 81, 88-90, 92, 110-114, 116, 123, 125, 128, 129, 133, 135

F

Fibre anisotropy 10, 18
Fibroblast 1-4, 7-11, 13-20, 25-30, 32-37, 40, 41, 48, 51, 53-55, 57-59, 61, 62, 64, 68-71, 74, 75, 77-82, 88, 89, 95, 96, 105, 111, 112, 121-125, 127, 129, 131, 133, 135
Fibroblast specific protein (FSP1) 51
Fibrocontractive disease 1, 4, 5
Fibrocyte 2, 40-45, 59, 77
Fibrosis 2, 4, 7, 25, 26, 32-35, 37, 40, 41, 43, 45, 47-51, 53-59, 61, 62, 64, 68-72, 91, 102-104, 106, 107, 110, 111, 113-116, 118, 123, 133, 135
FIZZ1 68-72
Force vector 7, 10, 11, 18-20, 125

G

Glomerulonephritis 47, 48, 50, 56

H

Hepatic stellate cell (HSC) 2, 88-99, 102-107
Hepatocyte growth factor/scatter factor (HGF/SF) 74, 79-81, 83
Hyaluronan 110, 111, 113

I

Idiopathic pulmonary fibrosis (IPF) 32, 35, 37, 113, 114
IL-4 33, 36, 68, 70, 71
IL-13 68, 70, 71
Inflammation 32-34, 36, 37, 40, 47, 50, 53-55, 69, 74, 102, 103, 110-114, 116, 121
Injury 7, 14, 18, 25, 32, 34, 40, 41, 48-50, 53, 54, 56, 68, 69, 71, 72, 91, 102-105, 110-116, 118-121, 123, 125-127, 129, 130, 133, 135
Innate immunity 110-117
Integrin 57, 81, 82, 97, 98, 106, 118, 125, 133, 135
Interferon-γ (IFN-γ) 33, 35, 110, 113, 115, 116
Invasion 57, 74-77, 79-83, 88, 90, 94, 96, 97, 99, 128

K

Keratocyte 118-125, 127, 128, 130-135
Knockout mice 48, 50

L

Liver metastasis 95, 97
Lung 10, 25-30, 32-38, 56, 68-72, 77, 82, 110-117
Lung fibroblast 25-30, 33, 35-37, 70
Lung fibrosis 33, 68, 113

M

Matrix 2-5, 7, 9-11, 13-22, 25-27, 32, 33, 40, 42, 45, 48, 54, 57, 68-70, 75, 80, 81, 88-93, 96, 102-107, 110, 111, 113, 115, 123, 125, 127-129, 133-135
Matrix metalloproteinase production (MMP) 18, 19, 57, 102, 103, 105, 106
Mesothelial cells 2, 53-55, 59, 61
Myofibroblast 1-5, 13, 15-17, 20, 21, 25-27, 30, 32, 33, 35, 36, 40-43, 45, 47, 54, 55, 57, 59, 68-72, 74-83, 88-99, 102-104, 107, 111, 114, 118, 119, 121, 127-136

N

N-cadherin 74, 82, 83, 105-107
Natural killer (NK) cells 91, 115

O

Orientation 7, 9, 10, 18, 119, 124-127

P

Peritoneal dialysis (PD) 53-62, 64
Peritoneal fibrosis 2, 54-58, 61, 62
Peritoneal membrane 53-55, 58, 60-62, 64
Peritoneal sclerosis 54
Platelet-derived growth factor (PDGF) 26, 36, 92, 91, 96, 118, 134, 135
Prevention of EMT 54, 62
Protease-activated receptor (PAR-1) 25-26
Pulmonary fibrosis 4, 25, 26, 32, 33, 35, 37, 68-72, 111, 113, 114

R

Remodeling 1, 4, 7, 13, 14, 16-18, 20, 22, 25, 26, 40, 41, 45, 48, 57, 68, 90, 92, 102
Resistin 69
Resolution 37, 102-107, 112, 118

S

Scarring 1, 2, 4, 7, 10-13, 18, 20, 22, 32, 33, 35, 82, 104, 107, 118
Scleroderma 25-27, 30
Shielding 21
Smooth muscle cell (SMC) 1, 2, 4, 5, 15, 25, 35, 42, 69, 75, 77, 80, 127
STAT6 68, 71
Stress fiber 2-5, 27, 37, 81, 82, 124, 126-128, 131
Stress-shielding 7-9, 22
Stroma 1, 4, 75, 80, 82, 88-90, 93, 95, 96, 99, 119, 121, 122, 125, 127, 129, 130, 132
Systemic sclerosis 1, 25, 26

T

Tenascin-C (Tn-C) 74, 79-83
Tension production 1, 3, 4
Therapeutic intervention 40, 54, 62
Thrombin 25-27
Thy-1 32-37
Tissue engineering 10, 11, 18
Tissue inhibitors of metalloproteinase (TIMP)
    57, 102-106
Tissue injury 25, 32, 34, 41, 91, 110-115
Tissue remodeling 1, 4, 7, 13, 20, 25, 40
Tissue repair 7, 13, 40, 58
Transdifferentiation 26, 54, 88-95, 97, 99
Transforming growth factor (TGF) 1-3, 16,
    20, 25-27, 35-37, 41-43, 45, 53, 55-58,
    60, 62, 63, 68, 71, 72, 74, 77-79, 81-83,
    96, 103, 105, 118, 130, 132-135
Tubular atrophy 47, 50
Tubulointerstitial fibrosis 48, 49

V

Vascular endothelial growth factor (VEGF)
    53, 54, 60-63, 82, 92-95, 97, 98
von Willebrand factor type C (VWC) domain
    29, 30

W

Wound contraction 4, 118, 120, 122-125,
    127
Wound healing 2, 4, 32-35, 37, 41, 45, 57,
    102, 103, 111, 114, 118-122, 124,
    127-129, 131, 135